Soldier Mountaineer

The Colonel who got Siachen Glacier for India

Soldier Mountaineer

The Colonel who got Siachen Glacier for India

Colonel N Kumar

with

Colonel N N Bhatia

Vij Books India Pvt Ltd

New Delhi (India)

Published by

Vij Books India Pvt Ltd
(Publishers, Distributors & Importers)
2/19, Ansari Road
Delhi – 110 002
Phones: 91-11-43596460, 91-11-47340674
Fax: 91-11-47340674
e-mail: vijbooks@rediffmail.com
web: www.vijbooks.com

First Published : 2016

ISBN: 978-93-85563-55-3

ISBN: 978-93-85563-56-0

Dedicated to

The Kumaon Regiment

WHY THIS BOOK?

I was fortunate enough to attend a United Service Institution function held in April 2010. This function was attended by all the three chiefs of the services where Col Kumar was awarded the McGregor Medal for his achievement 22 years back.

Later on I read an article written by Squadron Leader Rana TS Chhina (Retd) in the USI Journal of April – June 2010. I would like to add the first few paragraphs for the readers to understand the importance of the McGregor Medal.

> 'In the Nineteenth century, Great Britain and Tsarist Russia were the two major power blocs that influenced world affairs. In 1885, the Panjdeh Incident or Panjdeh Scare, rekindled British fears of a Russian threat to their Indian Empire through Afghanistan. Conflicting Russian and British interests in Central and South Asia for years had been the cause of a virtual cold war, known euphemistically as 'The Great Game'; and the Panjdeh Incident came close to triggering full-scale armed conflict between the two powers. Following the incident, the Anglo-Russian Boundary Commission was established to delineate the northern frontier of Afghanistan.
>
> One of the aspects that troubled the authorities in India was the lack of reliable information about the vast tracts of uncharted territory that lay along the remote and inaccessible frontiers of their Indian Empire. The person, who devoted his energies to filling this gap in his capacity as QMG and originator

of the Military Intelligence set-up in India, was Major General Sir Charles Metcalfe MacGregor, KCB, CSI, CIE, who founded the United Service Institution of India in 1870. Therefore, shortly after he passed away in February 1887, the USI Council instituted the MacGregor Memorial Medal (MMM) in May 1887 to commemorate his memory.

The criteria for award of the MMM were laid out at a meeting held on 03 July 1888 at Shimla, presided over by the C-in-C General Sir FS Roberts, with the Earl of Dufferin, the Viceroy, being present as Chief Guest.

Initially, the award was to be given only for significant military reconnaissance or journey of exploration or survey in remote areas of India, or in countries bordering, or under the jurisdiction of India, which produced new information of value for the defence of India. The award was usually of a silver medal, but a gold medal could be awarded in place of a silver medal or in addition to it, for especially valuable work. During the period of the British Raj, the MacGregor Medal became the de facto award of 'the Great Game' and among its recipients were names such as Sir Francis Younghusband (1890) and Major General Orde Wingate (1943). The first Indian soldier to get the award was Havildar (later Subedar, IOM) Ramzan Khan, 3rd Sikh Infantry, PFF (1891)'.

I have also read Col Kumar's book "KANCHENJUNGA" which is a story about the first successful expedition from the North East Spur. Some of the comments about this successful climb of Kanchenjunga from the North East Spur are appended below:-

> "We believe this one of the greatest achievements in mountaineering history…"
>
> –Dr. Eizaboro Nishibori - President, Japanese Alpine Club

> "Hearty congratulations. That was certainly one of the most difficult yet done."
>
> –H. Adams Carter - Editor, American Alpine Journal.

"Outstanding achievement by any standards".

–T.H. Braham- Swiss Alpine Club Journal.

"The brilliant effort was, I think, an even more notable achievement than nine atop Mount Everest"

–Soli S. Mehta - Editor, Himalayan Journal.

Later, I saw a documentary made by Films Division of India called "Guts & Dedication" which is about the achievement of Col Kumar. He is the only soldier mountaineer on whose life the Government of India made a film.

In October 1963, I have read the list of awardees on 26th January by the Government of India and saw that very seniors officers – Lieutenant General and equivalent being given the 'Ati Vishisht Seva Medal'. One of them is an Army Commander who has put in 35 years of services. I was surprised to know that this award was earned by Col Kumar as a Captain with 7 years of service.

I thought it was a high time somebody wrote a biography of Col Kumar who is perhaps the most highly decorated man of the services. Col Kumar was kind enough to give me the details of his exploits.

– Col N N Bhatia

Contents

Foreword

The bond between me and 'Bull' – the nick name Cadet Narender Kumar earned during the first boxing match he fought with his rival a senior cadet, Sunith Francis Rodrigues, who went on to become the Chief of the Army staff. Narender won the bout and earned the nickname 'Bull' for ever. His tendency to charge relentlessly into whatever he does, grew over our common passion for the mountaineering.

Bull was lured to mountains during his tenure in the KUMAON Regimental Centre by leading an expedition to Trishul (23,360 feet). Nothing thrilled him more than being on top of Trishul on 4 June 1958. His subsequent exploits put him among front rank mountaineers of Tenzing Norkey and Sir Edmond Hillary fame. He was a member of the first all Indian Expedition to Everest (1960) and Deputy Leader of the successful Indian Everest (1965) led by me that put nine Indian Army climbers at the top of the world's tallest peak, and the Leader of the Nanda Devi (highest Indian peak) Expeditions. He also led successful climbs to Chomolhari (highest peak in Bhutan), Nilkantha, Sia Kangri and Teram Kangari in the Eastern Karakoram ranges besides the Indo- German Boat Expedition and Trishul Ski Expedition. He was Principal of the Himalayan Mountaineering Institute, National Ski Institute and the Commandant of the High Altitude Warfare School. His exploits brought him many national and international honours and awards notably the Padma Shri, the Param Vishisht and the Ati Vishisht Sewa Medals, the Arjun Award, Fellowship of the Royal Geographical Society (FRGS), the Gold Medal of the Indian

Mountaineering Foundation and Mac Gregor Memorial medal. Bull's unique contribution towards the world of mountaineering was the ascent of Kanchenjunga from the North east spur. For 45 years, several expeditions tried to climb the Kanchenjunga peak from the dangerous Sikkim route and were unsuccessful. This achievement was considered so remarkable that he was awarded the Gold Medal at the Los Angles Olympics. Lord Hunt described Bull's achievement as being 'far greater than the Conquest of Everest as it involved technical climbing and objective hazards of a much higher order than those found on Everest.'

He was deservedly decorated with prestigious Mac Gregor Memorial medal in 2010, instituted by the United Services of India in 1888, in the memory of Major General Sir Charles Metcalfe Mac Gregor to recognize exemplary service in the fields of Military reconnaissance, expeditions, river rafting, world cruises, polar expeditions, running and trekking across the Himalayas and adventure flights. Col Kumar was honoured for leading, as earlier mentioned, multiple expeditions in the Siachen area of the Eastern Karakoram ranges, in uncharted territory, under extremely harsh weather conditions, with minimal equipment and administrative support and grave risk to life and limb during 1978 and 1981 and thus, gained highly valuable terrain and enemy information that was instrumental in safeguarding our borders in that area and subsequent launch of "Operation Meghdoot". In his personal capacity, Bull has led nine out of 13 expeditions to peaks above 24000 ft. He was also the first individual to cross Siachen from "Snout to the Source". He is prolific writer on adventure sports and has penned six books on mountaineering, skiing and rafting. A documentary film by the Film Division of India has also been made to honour him. He at this age serves the country as Associate Vice President of the Indian Olympic Association. Bull is an extremely passionate mountaineer with a missionary zeal as Almighty gave him wings to soar higher and higher on mountain tops as if for spiritual quest. With age, he has become humble, egoless to achieve great heights, yet, he bubbles with energy as he says 'if you are low in energy you won't love what you do' as while exploring mountains, every moment of the day is crucial to survive. In our country in these turbulent times Virat Kohli, MS Dhoni, Anna Hazare and Shah Rukh Khan are the top three role models. It is time to add to this illustrious list the name of Col Narender Kumar too.

Though I have never met Col NN Bhatia, I have been reading his articles published in numerous publications including the 'Turkey Tribune'. He is prolific writer on matters military and national security and has already authored two books. His biographic effort 'Soldier Mountaineer' on 'Bull' is most fascinating and absorbing, letting every reader feel being part of the 'Bull' entire adventure saga. I am sure it will inspire the future generations of soldiers and mountaineers world over.

Jai Hind.

Captain Manmohan Singh Kohli

Former President of the Indian Mountaineering Foundation & co-founder of the Himalayan Environment Trust

The philosophy of the Soldier Mountaineer Colonel Narinder 'Bull' Kumar!

The human spirit needs places where nature has not been rearranged by the hand of the man.

— *Anonymous*

Preface

The Regimental Centre Officers' Mess in Ranikhet provides the most grandeur view of the eternally majestic snow capped Himalayan peaks that inspired a young mountaineer Capt (later Col) Narinder 'Bull' Kumar of 3 KUMAON (Rifles) to conquer them. Bull was lured to mountains during his tenure in the KUMAON Regimental Centre by leading an expedition to Trishul (23,360 feet).

His subsequent exploits put him among front rank mountaineers of Tenzing Norgay and Sir Edmund Hillary fame. He was a member of the first all Indian Expedition to Everest (1960), Deputy Leader of the successful Indian Everest Expedition (1965) and Leader of the Nanda Devi (highest Indian peak) Expeditions. He also led successful climbs to Chomolhari (highest peak in Bhutan), Nilkantha, Sia Kangri and Teram Kangri in the Eastern Karakoram ranges besides the Indo - German Boat Expedition and Trishul Ski Expedition.

He was the Principal of the Himalayan Mountaineering Institute, National Ski Institute and the Commandant of the High Altitude Warfare School. His exploits brought him many national and international honours and awards notably the Padma Shri, the Param Vishisht Sewa Medal, Kirti Chakra and the Ati Vishisht Sewa Medal, the Arjuna Award, Fellowship of the Royal Geographical Society (FRGS), the Gold Medal of the Indian Mountaineering Foundation and Mac Gregor Memorial Medal.

Bull's unique contribution towards the world of mountaineering was the ascent of Kanchenjunga from the Northeast Spur. For 45 years, several expeditions tried to climb the Kanchenjunga peak from the dangerous Sikkim route and were unsuccessful. The leader of the German expedition who has attempted this route in 1931 was awarded the Gold Medal at the Los Angeles Olympics even though they did not reach the summit.

I had the unique privilege of meeting for the first time Capt (later Col) Narinder 'Bull' Kumar during investiture ceremony in the Rashtrapati Bhawan in Oct 1963. While he was being awarded Ati Vashisht Sewa Medal (AVSM) - the first Indian Army Captain to be awarded AVSM, my brother Major Prem Nath Bhatia (6 KUMAON) Hero of the 1962 Battle of Walong) was awarded Vir Chakra in the same investiture ceremony. I felt very proud when my brother introduced me to the legendary Capt Kumar as 'the renowned Indian mountaineer of the Regiment'. I found him down to earth humble, jovial and thorough professional. During our this very short meeting, he advised me to respect the mountains and the high altitude as my unit 13 KUMAON was then deployed in Darbuk after the world famous 1962 Battle of Rezang La. Whenever we meet, he always greets me by his nostalgic saying, 'when I meet you NN, I am always reminded of your great brother Prem'. My every interaction with him in the recent past has left an inedible mark on my mind.

Bull was decorated with prestigious Mac Gregor Memorial medal in 2010, instituted by the United Services of India to recognize exemplary service in the fields of Military reconnaissance, expeditions, river rafting, world cruises, polar expeditions, running and trekking across the Himalayas and adventure flights. He was honoured for leading multiple expeditions in the Siachen area of the Eastern Karakoram ranges, in uncharted territory, under extremely harsh weather conditions, with minimal equipment and administrative support and grave risk to life and limb and gain highly valuable terrain and enemy information that was instrumental in safeguarding our borders in that area and subsequent launch of "Operation Meghdoot".

He has led nine out of 13 expeditions to peaks above 24000 ft. He often proudly is mentioned as 'the Officer who got us Siachen'. He is also a prolific writer on adventure sports and has penned six books on mountaineering, skiing and rafting. A documentary film by the Film Division of India has also been made to honour him. He at this age serves the country as Associate Vice President of the Indian Olympic Association.

Bull is an extremely passionate mountaineer with a missionary zeal as Almighty gave him wings to soar higher and higher on mountain tops as if for spiritual quest. With age, he has become humble, egoless to achieve great heights, yet, he bubbles with energy as he says 'if you are low in energy you won't love what you do' as while exploring mountains, every moment of the day is crucial to survive.

In our country in these turbulent times, Sachin Tendulkar, Anna Hazare, Amitabh Bachchan and Shah Rukh Khan are the top three role models. It is time to add to this illustrious list the name of Col Narinder Kumar too. He has been an awesome soldier, trail-blazer, thinker, innovator, motivator and leader in the adventure sport.

It is a great honour for me to write his Biography.

– Col NN Bhatia

GENESIS

Narinder Kumar was born in Rawalpindi, now in Pakistan on the 8 December 1933. They were four brothers and two sisters. Narinder's father was an educationist and professor in the Chief College, Rawalpindi which was the alma mater for the royals of all the princely states like Kapurthala, Patiala, Chamba, Jind and other similar states in northern India.

The eldest brother, Brigadier KI Kumar was a great swimmer; and the services' champion in the breast stroke events. The next was Major Davinder Kumar who after 16 years of service in the Territorial Army (TA) Regiment went to Iraq to work for a construction company - Somnath Builders, around the time when Iraq had attacked Kuwait. One unfortunate night, he was ambushed by deserters of the Iraqi Army and asked to hand over the keys of the vehicle he was driving. He resisted and was shot by the deserters. Later on the offenders were caught and hanged.

Bull Kumar as he was christened in the Army – the actual story of how he was christened is explained in another chapter; was the third in the line of the siblings. The fourth brother was Major Kiran Kumar who was a Para Commando and had followed Bull's footsteps in mountaineering. Unfortunately, in 1985 while climbing the summit of the Everest, solo; he missed his footing and hurtled down to instantaneous death.

One sister, Saroj Pandit elder to Narinder, was married to the Indian National Trade Union Congress (INTUC) Secretary who later became the Minister of State in the Central Government. His youngest sister Nirmal Pandit was married to an Army Officer who retired as Major General.

Bull Kumar (we will continue referring to Col Narinder Kumar as Bull Kumar or just Bull) was married to Mridula Sadgopal on the 21 February 1966. Mridula was doing a course in perfumery in Sweden at that time. Most of the Indian 1965 Everest team members got married after the successful Expedition. Mridula's father was the topper from the Banaras Hindu University in his batch. He had joined Indian Standard Institution and dealt with basic oils. Her mother Kamla Sadgopal was born in Lahore and was one of the first few ladies to complete her B.A. in her time.

His son Akshay Kumar was a great skier and would have partaken in the 1988 Olympics but when he was sent to France for training, he fractured his knee unfortunately; serious enough to bring an end to competitive skiing. He had studied in the Delhi's prestigious Modern School and subsequently graduated from the elite St Stephen's College. After that he joined Mercury Himalayan Explorations, known for organizing treks, rafting and other adventure expeditions. Since he was also a great rafter, he led the Brahmaputra Rafting Expedition form Tuting to Bangladesh. He has also done Narmada and Ganga Rivers' Rafting Expeditions from their sources to sea. He presently holds the eminent positions of the Chief Executive Officer (CEO) of the Mercury Himalayan Explorations and also the President of the Adventure Tour Operators Association of India. Akshay's wife Dilshad Master was the Vice President of the Star TV. They have been blessed with one daughter.

Bull Kumar's daughter Shailaja Kumar was the first lady to represent India in 1988 Winter Olympics held at Calgary. She is married to Yash Karan Singh and is blessed with a daughter. They live in San Diego, California.

Bull's father being a teacher correctly believed that young children should not go to school at an early age. Bull was 8 years old when he started going to school in Class III and had only 7 years of formal education before he joined the Joint Services Wing (JSW) at Clement

Town, Dehradun. This institution was later known as the National Defence Academy (NDA) and shifted to Khadakvasla near Pune.

School days were fun as he was a Boy Scout. The scouts used to be called down to Annandale Race Course where they used to act as staff members to manage the crowds. At the end of the races, the management used to give them a sumptuous and rich tea. It was there that he learnt to bet on horses.

Scouts used to pool in money and the jockeys and owners gave them good tips to enable them to make a little extra pocket money. After that they used to climb up from Annandale to Shimla. They were young and fit and sometimes they climbed upto Jakhu - the Hanuman Temple, where there used to be lots of monkeys and return in the evening.

In those days climbing 3,000 feet was not considered to be a great feat. From Shimla there was a road that led to Tibet. Once some of the boys got bored and decided to go to Tibet. They carried a few puncture units with them and thought that they could survive on *gur* (jaggery) and *mungphali* (ground nuts), but they got far too many punctured tyres in their bicycles en route! Their Tibet mission remained a dream, but later on, on many occasions and in many sectors, Bull Kumar entered Tibet during his numerous mountaineering expeditions. He visited Barahoti in the Central Himalayas in 1961, Lola Pass in 1965, conducted an Everest Expedition in the Eastern Himalayas and the 1981 Siachen Expedition in the Western Himalayas, shattering the myth that dreams do not really come true.

Bull studied in the Government High School, Shimla, and used to play cricket matches with Bishop Cotton School in Shimla. They won many a match! They usually played in their ground and were looked after very well and given good lunches and tea. Once the Bishop Cotton School team came to play in Government High School ground but the hospitality shown to the Cottonians left much to be desired!

All the brothers were extremely lucky that their father was posted as the Headmaster of the Government High School, Shimla in 1946; a year before the partition. After the partition, their small house had

50-60 guests at a time, as most of the refugee families were devastated due to partition of the country. All the young people were lodged in the boarding school while the ladies and gents slept in separate rooms. They shared the meagre resources of the family. The boarding school children had the time of their lives during those trying days which forged a permanent bond till date.

While Bull was in Class IX in 1947 he went to World Scouts Jamboree in Paris. There he mingled with International scouts from all over the world. In the evening every country had to put up their own cultural shows at the camp fire and he kept visiting separate countries every day. The Indian contingent consisted of 150 boy scouts out of that 50 were from Punjab itself. Mr Thaddeus from the South was the Contingent Commander and Mr Qureshi was the head of the Punjab Contingent.

On the eve of 15 August 1947, they hoisted an improvised Indian and Pakistani flags together with the Union Jack in the centre. It was only while sailing back from Paris in the ship called 'Strathmore', when one night the churning of the engine stopped and their Pakistani friends including his best friend Sarfraz were taken away to Karachi while the rest journeyed to Bombay. It was only at Bombay that they understood the havoc that the partition had brought about in the Motherland. Thousands of migrants, both Hindus as well as Muslims were being slaughtered on either side of the border.

At Delhi Junction they all parted company. Even the Punjab Contingent broke off to different areas. Some went to Amritsar, Shimla and to other cities of Punjab. He still remembers a scene where a young boy named Swaran Singh who came from a very rich, affluent and cultured family from the Western Punjab found his parents penniless on the railway station after the partition. At that time, he was too young to understand such privations.

He also remembers that before going to the Jamboree, he needed a full sleeve sweater and four of his aunts knitted a sweater in a combined effort. When he went to the JSW, he was asked to give a small talk during ragging and he started talking about the jamboree he had attended and earned the nickname "Jamboree".

The first JSW course had only an interview and no examination. He got selected in the interview and was very confident to go to the Academy. All of them, who were selected, stopped studying, sold their books and waited for the call from the JSW. At the last moment, it was announced that there would be a Union Public Service Commission (UPSC) written examination. Firstly, he thought that was a mere formality but later on he realized that it was a competitive examination and he was not in the order of merit. So, he joined the JSW a year later in 1950.

Cadets stayed in the barracks and ragging was very much part of the training at that time. He personally felt that healthy ragging was really essential to break down one's ego. Putty parades were another custom that could also be given to junior cadets by the senior appointments. Putty parades included asking juniors to go and change their dress in three minutes time or report in a particular dress in the evening where they made them do front and back rolls or run around the barracks; and no cadet could escape these punishments.

There was another type of punishment called extra drills that could be given by officers if cadets were caught slouching; or breaking any custom or tradition. Apart from extra drills they were also given restrictions debarring them from going to the town on out passes on Sundays and holidays. The restriction drills also meant reporting at night where some officers would take 30-45 minutes' parade. There were some cadets who were perpetually on restrictions and would take life easy as they could not do more than one restriction drill in a day.

Some of them had joined JSW to escape studies but were surprised when they had to take at least 17 subjects that included academics and military studies. Each room had four cadets, all of different terms. In his case, there were first, second, third and fourth termers! It worked both ways; the good point was that the senior cadets could help juniors in their work and guide them on how to overcome daily difficulties, while most of the times the junior cadets had to run errands for the seniors.

Once while he was bringing a chocolate for his senior, enroute he was accosted by another senior cadet who demanded a portion of the chocolate and he could not refuse to give him some. Since then, the senior cadet decided to buy his own chocolates lest they be swiped en route.

Since malaria was prevalent in those times, anti-malaria precautions were mandatory and cadets had to roll down their sleeves in the evenings and the mosquito nets were drawn to sleep. Even now he seldom finds an army officer with half sleeves in the evening. Somehow it had become a habit with him too.

There used to be two Mess Nights a week where they had to go formally dressed and had a sit down dinner. After the dinner, there was a custom to drink a toast to the President once everybody had finished their meals. The President would rise up and say, "Mr. Vice for the President" and Mr. Vice would then get up and say, "Gentlemen, the President" and everybody would rise up and say, "The President". Before independence the toast used to be drunk to the King. This tradition is still in vogue in all the Armed Forces Officers' Messes.

The day's routine started with physical training (PT) in the morning, followed by breakfast, drill, weapon training (WT), academics and studies on the basics of the various military subjects. Lunch was followed by the compulsory siesta and games in the evening and the day ended with the formal dinner. After that there would be lights out time and everybody had to go to sleep.

Cadets were shown two movies in a week. They used to have Saturdays and Wednesdays as half days and Sundays as the holidays. They could go to town but had to be properly dressed and some of the areas and cinema halls used to be 'out of bounds'. Howsoever rich any cadet may be, they were not allowed to spend more than Rs.40/- a month. Even the children of Rajas and Maharajas had to stick to that rule.

There were various games to play and boxing was compulsory in the initial stages. Any cadet could pick boxing as a sport if he desired to. There used to be inter-squadron competitions and those who did well in sports used to get '*Blues*' in various sports. Bull Kumar earned his boxing '*Blue*' very early.

Only in riding they used to award '*Canes*' for those who excelled in riding, instead of '*Blues*'. Cadets could wear the 'Blues' on their blazers and got extra marks on their passing out merit list for sports activities. In fact sports and other extra curricular activities counted more than studies. Narinder remembers he passed out within top 10 percent of the passing out cadets whereas those who were very good only in studies were left far behind.

The cadets who passed out went to different academies. The Army cadets went to Indian Military Academy (IMA), Dehradun and the Naval and Air Force cadets went to their respective academies before they got commissioned.

During the passing out, as a custom, the junior cadets had the chance to rag their seniors. Bull still remembers some instances of his stay at JSW. One was that he used to study after lights out with a torch under his blanket. He had also represented his squadron in 5-6 games that left him quite exhausted at the end of the day.

In another incident before passing out, he was pinned down by 4-5 cadets and half of his moustache was shaved off. He was extremely angry and was out to take revenge and the alleged cadets hid themselves in the kitchens and never slept in their beds. On the lighter side, one of them had many years later written a column in the newspaper produced below:-

Moustachioed musketeers

By Bilu Suri

WHY is it that a person who had been sporting a moustache looks like a helpless, shorn lamb when the upper lip is shaved. Far more explicit, analogies can be given to describe this state but editors may be averse to publishing them.

The Fourth Course of the Joint Services Wing, in Clement Town, Dehra Dun, passed out in the summer of 52 with fully clean faces, except, of course, the Sikh cadets and one exception. The only exception permitted was Iqbal Singh who belonged to an old Rajput family. He pleaded that his father would take the removal of hair from the upper lip as a personal affront since their customs decreed that this step was resorted to only when the father dies.

It was a few days before the Passing Out Parade that Satish Malhotra, Suresh Chitale, Parmarthi Raina and I were sitting down peacefully after dinner reminiscing about the two years that appeared to have sped away, when all of a sudden the other three appeared to have gone berserk. Two pinned me down, and the third whipped out a razor, and very soon I was bereft of a virgin moustache. The first two being Naval Cadets were not permitted hirsute upper lips, and Parmarthi felt that unnecessary hair would interfere with the piloting of the latest fighter that he was aspiring to take on.

The indignity of losing a well-cultivated moustache could not be suffered in loneliness and hence the launching of Operation Clean Upper Lip. It was decided that no cadet of the 4th JSW Course would attend the Passing Out Parade with even a downy upper lip.

The Operation progressed successfully and we found that the most time saving and comical method was to shave only half of the moustache, leaving it to the victim to complete the rest of the operation. We had, however, not thought of Bull (now Col) N. Kumar, who later conquered the Everest and acquired a string of awards. It required about six of us to hold him down and with difficulty we could only clean a quarter of his upper lip. He swore that even though he would Pass Out without a moustache, Bilu Suri would do so also bereft of his eyebrows. All of us who have known Bull's tenacity were certain that this was not an empty threat.

There were three days left before Passing Out, and these were spent by me in the Cadets' Mess looked after by the Waiters — friendships formed during the last two years came as a blessing and I was saved, even though soiled table sheets used as bedding have a peculiar smell that can turn off even the most seasoned nostrils.

Photographic proof is available to authenticate this report for those whose memories have started to fade during the 45 years that have lapsed.

In those days hockey was far more popular than cricket. Narinder made his mark in cricket but then later on he realized that the matches took place during Saturdays and Sundays with the result his 'weekends' went for a six. It was very easy to play a couple of bad shots and get out of that game. The new games that he had picked up during the JSW were cycle polo and under water swimming. Cadets could also propose outdoor excursions like trekking, shikar, cycle trips, and the like. Once they were taken on a cycle trip from Clement Town to Mussoorie and he was the only cadet who did not get off the cycle. His squadron leader Atkinson was forced to keep pace with him and told him later that he had taken five years of his life that day.

RAFTING - HIS FORTE

Bull Kumar had a love for mountaineering at a very young age. Once, he remembers - four cadets decided to go to Mansarover Lake, but he opted out as he wanted to attempt Mount Kailash – which was day dreaming indeed!

It was from one of the excursions in the Academy that he went to Khalsi on the bank of the Yamuna River on a riding trip and there he saw lots of rafts passing by. Himalayan Rivers are fast. Water rushes down the Himalayan slopes with tremendous force, and with the momentum it gathers, it moves though the valleys at extremely high speeds. These rivers are dangerous. Because rocks and boulders lie on river beds and water flows over and around these. In addition, there are strong, unpredictable currents, whirlpools, rapids and waterfalls.

These rivers are fascinating because they are fast, dangerous and beautiful. The clear blue water and white foam bordered by sheer rock canyons, occasional pebble beaches or vast mountain ranges give to the Himalayan rivers a wild and strange beauty.

He was only seventeen when he first felt attracted to rivers and what began as a pastime developed into a major pre-occupation. During training as a cadet at the IMA, greater emphasis was given to physical development. The cadets had, in fact, become so used to physical exercises that even Sundays and other holidays normally meant to be days of rest for their tired bodies, found them eagerly on the move, and restless if they were not doing some physical activity.

One weekend trip they went horse riding to Mandi, twenty-six miles away, and were to camp there overnight. Before settling down for the night they massaged their horses only to find, much to their annoyance, that they, instead, smelled dirty. The Yamuna flowed close by, so early in the morning they decided to bathe in the river.

They all walked up to the Yamuna but only Bull Kumar stripped and was about to step in, when suddenly he stopped. There, before him flowing by majestically on the waves of the river was a huge wooden raft, forty feet long and twenty feet wide. Navigating it, as it rose and fell with the rolling motion of the waves, was a local boat with a boatman using only a long pole. Narinder's heart was beating faster as he thought how he would love to be on that raft. His companions apparently felt as he did, for their faces showed similar emotions. They too were excited by that sight.

They began waving their arms, signaling frantically for the man to stop, but he only waved back at them and went past. That left them disappointed, but even more determined to ride a raft, right at that moment! The question was how? A plan was laid with great ingenuity quite reflective of their military training, and promptly carried out. All the currency notes that they had among themselves were slung together as a garland. The garland was then strung to one end of the long pole. That completed the preparations, after that they sat and waited.

Soon another raft came floating by. When it was close enough, the pole with its garland of currency notes was stretched across the river, so that the boatman would see it. He did, and as they had expected, he was tempted by the sight of all that money. He swung his raft around and steered it towards them. He landed close to where they were standing, waiting for him. They bargained with him and a deal was struck. They would pay him fifty rupees for a long trip down the river.

Though skeptical of the whole idea, their officer accompanied them, reacting to their enthusiasm and excitement, permitted them to undertake that adventure. And soon enough, preparations completed, tinned food stuff packed in a basket to keep them going till lunch time, the seven of them, all excited and shouting, boarded the raft. A massive shove by the navigator with his pole, and they were off. Narinder was delighted to be in the midst of such an adventure. Little did he know then how (mis)adventurous that particular ride would turn out to be.

As the raft moved downstream, bobbing along on the swift flowing waters of the Yamuna, they felt marvellous. They were soon used to the peculiar rocking rhythm of the flowing raft. Taking turns, each one of them tried navigating the raft with the long pole. None of them, however, succeeded. The raft man only laughed at their futile attempts. But that did not matter to them. Nothing mattered, except the thrill of the experience.

Talking, shouting, laughing, caught in the euphoria of the moment, none of them, not even the raft man, noticed the small cluster of rocks towards which the raft was heading. When they did, it was too late. The raft man tried, but could not prevent the raft from hitting straight into the rocks with an astounding force. The impact, as it struck was powerful, and they were momentarily stunned.

Barely had they recovered from the first shock that they received a second one. One end of the raft was hopelessly, immovably wedged between two rocks. They pushed and pulled, and tried hard; but it was so thoroughly jammed that it would not budge. Finally, they gave up trying. There was only one way out. To cut the ropes that held together the logs of the raft, at a point and break the raft into two halves, and then to float away on the half that would thus be freed, leaving behind the other stuck among the rocks as it was. The raft man agreed to that, since that was perhaps the best option, the only other being to abandon the raft and swim ashore, that none of them were keen to attempt!

The brilliant plan was executed perfectly - well, almost perfectly! Everything went fine until the last rope was cut out, and one half of the raft was free to move. Move it did, but so quickly, captured as it was by the powerful, swift current, that only the raft man was able to jump aboard. He tried desperately to swing around, and come back for the rest, but the current was too strong. Quite disbelievingly they saw the raft and the raft man being carried further and further away, receding into the hazy distance. Now they were really up the river with half a stuck raft and no paddle!

To release the tension they laughed, as one of them shouted after him to at least take his money. But he was already too far away to hear them, and even if he did, could do nothing about that. Not only had he left behind the money they owed him, but also half of his raft. There they were, stranded, with no way out. At least that is what it appeared at that time!

To the young red blooded cadets, it was even now a heady adventure. Their officer, Captain Dhingra, however, was worried, and with good reason! While he had allowed them to step into that venture, swayed as he was by their enthusiasm, desire for fun and excitement, he also had the responsibility of getting cadets out of that tricky situation. So, at his urging, the cadets sat down to consider various options.

Their best option seemed to be somehow dislodging the remaining half of the raft. So they tried once again. At that time their determined and concentrated efforts succeeded. To their exuberant shouts of triumph, the raft gradually came loose. They were quick this time in jumping onto the raft, before it could be carried away.

Hardly had they settled down on this new phase of their adventure when the navigational device, the pole was missed. The raft man had, of course, taken it away with him. For a while cadets were nonplussed, but only for a while for they remembered their earlier, and futile, efforts to navigate the raft with a pole. Only an experienced navigator could successfully have used it, not the novices. So they relaxed and settled down once again to enjoy themselves.

Furiously and almost rhythmically, the waves carried them on. Narinder looked at the long line of trees along the river bank. The trees seemed to be moving backwards, rushing past his eyes, giving him a pleasant sensation of speed. The sun was shining down hard upon them, scorching their naked skins as they had took off their clothes and were down to their undies. Sometimes, a wave larger than the other would glide over the raft, icily splashing Bull's sun burnt skin, cooling and soothing it.

For a while the cadets went on that way, enjoying the speedy sensation of a water ride at a park. Then one of the cadets, hungry perhaps, or merely wanting to munch, asked for biscuits. A cry went up as it was discovered that the lunch box was missing. It had been carried away on the other half of the raft! Their initial disappointment, however, did not last long. Despite hunger gnawing at their innards, they were happy that the raft man had at least partially been compensated, since their basket had been really well stocked. To forget their hunger, they sat down to sing community songs that had been taught to them at the Academy.

Soon they became thoughtful once again. It was only then that the grimness of their predicament hit them. They were hungry, but had no food. They were on a raft in the middle of Yamuna, totally at the mercy of its waves, but had no means of navigation. There seemed no way of landing on the banks. To any question regarding the outcome of that foolhardy adventure, there was perhaps only one answer: "God knew only what would happen!"

As an old saying goes, "Man's extremity is God's opportunity" And that is exactly how it turned out to be, for in the distance they sighted the dome of a Gurudwara-probably Ponta Sahib. As they were going to pass through a town, they might possibly be able to land. They still had no idea as to how they would, but they said their prayers anyway.

As the raft neared the Ghats, they shouted out, *"Bachao, bachao"*. Those who did hear them were, however, helpless, for the raft was too far out in the river for them to do anything except see them racing past at a very fast pace. As the raft crossed the Ghats and the Gurudwara Ponta Sahib, the cadets saw that they were once again heading straight towards a huge rock. The current was awfully strong, perhaps even stronger now. There was no way of avoiding collision, so they braced themselves and waited for the inevitable impact.

With a loud retort and immense force, the raft rammed into the rock. Bull, along with some others, was thrown forward onto the rock. A few were, however, thrown into the water. Quickly they scrambled up to the relative safety of the rock. On impact, the bundles containing their clothes, shoes, and money had been thrown into the river. They then watched helplessly as those were carried along on the waves for some distance, only to sink. Only one cadet had been able to grab and retrieve a bundle. He was quite thrilled at his achievement.

Again they were stranded. Looking around, Bull Kumar saw that the bank at that particular point was quite close. He also looked at the raft, damaged but floating as the powerful pressure of the waves kept it pressed against the rock. To him it seemed pointless to cast on once again. Not only would there be the original problem that of their inability to steer it towards the bank, but there would be an additional danger too. The already damaged raft might break up midstream.

The time had come to act, he thought, and without even waiting to consult Captain Dhingra, he chose to follow the only course open to him. With a mighty shout of 'abandon ship', he dived into the swirling waters. Once in the river, swimming towards the shore, he felt free of the false sense of security that had taken hold of him on the raft. He felt free to make to the shore by his own efforts, or to drown, trying. The waves were powerful, but so was Bull, a powerful swimmer. Slowly and surely, he swam towards the bank.

Back on the rock, the others were stunned by the sudden turn things had taken. The cadets looked to Captain Dhingra for some indication as to their next move. He let out a shout, 'Stop, Narinder, stop!' But he was too far gone to change direction. Anyway, he did not want to go back to that rock. So he carried on swimming away from them.

Captain Dhingra had dived into the water and swam after him, as if on cue, the other cadets also jumped in the water, one after the other to save him. Like Bull, all of them were good swimmers. Swimming powerfully, all of them made it to the bank. Once there, safe and happy for the moment, they relaxed. Looking back at the rock, they saw the raft slide back into the water, and float away, damaged though it was. This then, was Narinder's first experience with rafts, and rafting.

Thus they were back on the shore, wearing only thin cotton underwear, wet and sticky. The cadet who had saved the bundle opened that only to discover that it had Narinder's clothes and not his that he had brought back to shore. Narinder gratefully accepted them, ignoring his good natured cursing as he put them on, while the others stood around, somewhat embarrassed at the close scrutiny they were subjected to from those who had gathered to witness the landing.

It was two in the afternoon; they were without money, starving, bone tired and weary! Understandably so, Captain Dhingra, the only Sikh among them, suggested that they visit Ponta Sahib, the Gurudwara they had earlier seen, and eat at the langar. The suggestion was gratefully accepted. Cadets were welcomed at the Gurudwara by the Giyani Ji, and they ate a simple, delicious lunch of dal and rotis. After lunch, they walked to the bus stop, two kilometers away, and boarded a bus back to their camp. Narinder's companions once again faced surprised, if somewhat amused, looks from the other passengers in the bus.

Finally they were back in camp. Their arrival was greeted by caustic remarks about their appearance. There was general fun and laughter as they related their experiences. Bull was sure however, that everyone who had not shared in their adventure must have wished that they had. It really had been an eventful, exciting day. Anyway, at the end of it all, they triumphantly rode their horses back to the Academy.

IMA - CREME DE LA CRÈME!

Life at the Academy went on as usual. They never did rafting again. But in those days Bull Kumar often thought of that wonderful experience - his first trip on a raft, and he was sure even then that he would do it again someday.

The IMA training of two years was overall a great education. He continued his cycle polo and Captain Thakur Kishan Singh – a great polo player from Rajasthan used to be always in the rear giving long shots and he used to dash with a cycle to score the goal. They used to have bets wherein the losing team had to pay the bill for the party that generally followed after the game - the venue being the famous Quality Restaurant in Dehradun. Captain Thakur Kishan Singh had a Rolls Royce and they would all go in the car to enjoy the parties. The winning team could take the maximum advantage by ordering anything and sometimes in their exuberance they over ate and suffered torture from a rebellious alimentary canal! However, when you're young, you could hog and enjoy anything!

From cycle polo, Bull graduated to playing polo on horseback. Polo is a team sport in which players scored by driving a small white plastic or wooden ball into the opposing team's goal using a long-handled mallet.

As Bull Kumar was a strong rider, he used to be given a good horse and he had the lowest handicap at minus 2. He used to be told to check the best player of the opposite team and he rarely took a shot at the ball. He remembers in one match, he was asked to mark Rao Raja

Hanut Singh of Rajasthan who had handicap of 9. He used to ride him off whenever he was about to score a goal so that he could not hit the ball. Once he got so mad that he even lifted his stick to hit Bull, but luckily better sense prevailed.

Apart from polo, the cadets sometimes took part in pig sticking which is a very dangerous sport. Pigsticking is a form of boar hunting done by groups of spearmen on horseback using a specialized boar spear. In India, pigsticking was popular among the Jats, Gujjars, Rajputs, Sikhs, Maharajas and the British officers during the Victorian and the Edwardian times. According to the 1911 edition of the *Encyclopaedia Britannica*, it was encouraged by military authorities as good training because "a startled or angry wild boar is a desperate fighter (and therefore) the pig-sticker must possess a good eye, a steady hand, a firm seat, a cool head and a courageous heart."

When men go on hunting the ladies followed them on an elephant back and in the evening participants usually camped at a nice spot and the ladies joined them and a big party was organised.

His mentor in riding used to be Captain Thakur Kishan Singh and once after a chase, his horse stumbled into a ditch and he lost his spear. Coming back to the camp without one's spear was considered shameful especially for a great rider like him. So, he took Narinder's spear and the latter being the novice, nobody noticed that his spear was missing.

Another horse sport that cadets took part was the fox hunt. Fox hunting is an activity involving the tracking, chase, and sometimes killing of a fox, traditionally a red fox, by trained foxhounds or other scent hounds, and a group of unarmed followers led by the master of foxhounds, who follow the hounds on foot or on horseback.

Apart from hunting and playing polo, Bull Kumar took active part in boxing, hockey and swimming. As a cadet, one could box other ranks and once he had a very close fight with the then National Champion who was from Gurkha Brigade stationed in Dehradun. Another big boxing bout he had was with cadet SFA Rodricks who had been trained in boxing earlier in school. He was 6 inches taller than him plus had a very long reach. In the first two rounds he had thrashed Narinder who kept on charging on him and Narinder finally knocked him out. The Commandant of the IMA at that time said that Narinder fought like a bull and so the junior officers followed and since then Narinder was nicknamed the "Bull". His old pet name "Jamboree" remained with his old friends from the JSW.

Throughout his stay at the Academy, he was on the warning list for relegation due to his being weak in studies. The only plus point he had in studies was that he was very good in mathematics. Once he got 97% and when he questioned why his 3 marks had been cut, the civilian teacher simply said, "bad handwriting".

It did not leave him satisfied. Another subject where he got good marks was drawing. It was not because that he was good at that, but their drawing teacher was also in charge of indoor games and Bull Kumar used to be the chess champion and the teacher always gave him cent percent marks.

Even after passing out, many years later when he went to the IMA to give lectures about his mountaineering expeditions, he was really scared to pass through the Adjutant's Office.

In the Academy days one of the tests was drill square and unless a cadet passed his drill square, he could not go on out pass to town.

When the time came for passing out, he opted for the Para Regiment. At that time Lieutenant Colonel TR Jaitley, his maternal uncle was commanding 2 Para. But somehow, he never felt comfortable in joining the Battalion where his uncle was the Commanding Officer (CO) because for the rest of his life he would have carried the tag of "*Mama Bhanja*". Hence, he joined 3 Kumaon Rifles.

The reason why he joined the 3 Kumaon Rifles was that one of the Battalion Officers Captain KS Kataria was the Adjutant who used to carry his cane the other way round. The knob that should normally face the front was carried behind, and the story behind it emerged that when one of the Commanding Officers went to give the report to Brigade Commander that the Battalion was ready for inspection, he inadvertently carried his cane other way round. The other Battalion officers followed suit and since then the tradition has been kept. So, he thought that 3 Kumaon Rifles must be a Special Battalion!

When he was passed out, the Commandant of the Academy Major General Habibullah wanted him to join the Armoured Corps as he used to play polo. But in those days most of the officers who joined the Armoured Corps came from the princely families, like the son of the Raja of Kapurthala and the son of Maharaja of Jaipur popularly known as "Bubbles". Coming from an average family, he thought that he would be handicapped working together with the princes and that was the reason he chose the Infantry over the Armoured Corps.

All newly commissioned infantry officers who got posted to their respective battalions in the various regiments had to spend 21 days in their respective regimental centers. 2/Lieutenant Gurbax Singh and 2/ Lieutenant Afsir Karim were the other two officers commissioned with Bull Kumar and reported at the Kumaon Regimental Centre, Ranikhet. Later in life, it was no major surprise for Bull Kumar to notice how one's career could change as he adapted himself to the prevailing and changing environment.

Afsir Karim transformed himself and proved to be an excellent officer and become one of the fifteen Major Generals their course had produced. After retirement, he has often been giving talks on television on national security and has written many books on matters military.

After Kumar's short stay at Ranikhet, he had to join his Battalion -3 Kumaon (Rifles) at Tangdhar Valley across the Nasthachun Pass to be crossed on foot. The life of a young officer (YO) was not an easy one in those days. There was a senior subaltern and then for the first few years, the YO was not called by his rank, and addressed as "Mister". YO's were made to do weapon training (WT), physical training (PT) and field craft cadres with their jawans and also take classes for troops.

Bull Kumar was once detailed to give a lecture to his jawans on the handling of the hand grenade. In order to make his lecture more interesting and easy to understand, he picked up a hand grenade and took it to a blacksmith. He got a hole punctured in the grenade, emptied all the gun powder out and then got that dissected into two parts. All the time he was awfully scared that the grenade might explode. So, he kept pouring water on the grenade while the metal cutting process was being done to keep it cool. His effort as the junior most officer was highly appreciated both by officers and men of 3 Kumaon (Rifles).

The junior officer even in those days was treated as non entity in the Officers' Mess. Young Officers or YO's were to be seen and not heard! They could not turn on the radio unless he was a senior subaltern. So sometimes to save the situation, YO had to do the filthiest jobs whether he liked or wanted to do them or not. Once there was a big mess party in the officers' mess and he was in charge of seeing that the bathrooms were always kept clean. In those days they never had a flush system and piss pots were in vogue and if the sweeper missed them, the YO had to empty the piss pots himself.

Bull Kumar was a youngster and had never played bridge in his life. Once he was asked to join in to make a foursome on a bridge table and his CO Lieutenant Colonel SL Basera took him as his partner. He did not know the rules, though he had a very strong hand with Aces, couple of Kings and Queens yet he said 'pass' when his turn to bid came.

When his hand was seen by the others, the post mortem started and his CO asked him why he had not bid. He told him that he did not have a five card suit. The CO exasperatedly remarked that, did he need a Barthia (A formal Army dress) instead of suit to bid? Everybody laughed and Bull Kumar was fearfully embarrassed.

So, the next time he went to Srinagar, he found a book on 'How to play Bridge' in one of the book shops. He bought the book and learnt playing bridge solo while he was alone on the pickets holding the cease fire line.

Later when he was posted at the Geeta Post in Tangdhar, he came to the Battalion Headquarters for some work. He was again co-opted in bridge foursome and that time the CO was in the opposite side. By then he had learnt good amount of bridge and during his bidding chance, he bid a little slam that meant that he had to make all the tricks except one. His CO doubting Bull Kumar's capabilities doubled, but

Bull Kumar redoubled and made his little slam. It was another matter, that later he was taken to task by his senior subaltern for daring to redouble his CO.

He had to learn immensely as a young army officer. Bridge is a very addictive game. He remembers when he was posted in the Army Headquarters, New Delhi they used to have a one hour lunch break. The Queen Victoria Officers' Mess was next to their offices in South Block. They came quickly to have a bite and played couple of rubbers of bridge before joining office again. He still remembers that later on when he was the Principal of the HMI Darjeeling, there was a bridge championship where in he and the Deputy Commissioner of Darjeeling Mr TS Brocca were partners and had won that championship.

He had to go for a physical training (PT) course to Pune and one PT Havildar was made in charge of his pre-course training. During that period in the officers' mess there was a big party and after a few drinks, the Training Officer (TO) joked if he was prepared for the PT Course! He replied, 'Yes Sir' and then the TO asked him to do a 'hand spring'. Though he was not warmed up, he still managed to do the perfect hand spring but in the process hurt his back badly.

The Regimental Medical Officer (RMO) advised him not to go on the PT Course but he replied that missing the first course would be taken as shamming. Due to backache he was unable to get up in the mornings and used to ask his orderly to warm up his back with an Infra-red heating device and only after that he would go and do the pre-course physical training. They used to get afternoon break but he could not rest as his back would get stiff again. He got the highest grading 'AX' in the course and later went back for the Advance PT Course.

With him on the Advance PT Course was Lieutenant Zaki who later on became Commander of IPKF in Sri Lanka. He was an extremely upright officer, outspoken, very intelligent and an excellent commander. While doing the course, once they were joined by a squad of other ranks who were being trained only for boxing. The two Havildar instructors of both the squads decided that there should be a boxing match between the two squads. Lieutenant Zaki told the instructors that it was forbidden for officers to fight other ranks in boxing. But the boxing squad Havaldar made some derogatory remarks that the officers should put on bangles. Bull Kumar could not take that insult and he decided that he would take on one of the soldier boxers.

His opponent was a very tough looking Lance Naik. They had chosen the best boxer from their squad. However, boxing needs lots of stamina and training and he had not been practicing for any bout. So, he made a condition that they would fight only for one round. Both the instructors agreed. In those days each round was for two minutes. They got into the ring, shook hands and started boxing. Bull Kumar thrashed the hell out of his opponent, getting him to the floor twice. Though no points were given, he must have won by a large margin. After the fight he stepped out of the ring.

The havaldar instructor of the other squad asked for the second round, but he refused. That was the talk of the town and all the officers were very proud of Bull Kumar as he kept the 'IZZAT' of the Officers' class. Bull Kumar recalls, in the Academy, he was always known as a fierce boxer and many of the opponents used to offer him temptations so that he would hammer them less.

In those days, Bull Kumar was very active in sports. Besides being boxing champion, he played polo, hockey for the Company and took part in swimming. He had consulted a doctor who advised him to take it easy in sports. He kept playing polo and decided not to take part in boxing any longer as it was a very exhausting sport and not favoured by the Officers' class.

In 1957, he was posted in Tangdhar with his Battalion 3 Kumaon (Rifles). When Tangdhar Brigade Headquarters gave his unit vacancy for the Winter Warfare Course, he was detailed for the course - the toughest course of the Indian Army. They were kept in the Army Transit Camp in Srinagar for 21 days and every day they were taken on a road walk and run on top of the hill to the Shankracharya Temple. So, after toughening up they reported to Gulmarg Winter Warfare School.

WINTER WARFARE SCHOOL

The Winter Warfare Course was started by General KS Thimayya, DSO. His ski instructors were Lieutenant Dhillon from the artillery and Captain Prem Hoon who later retired as an Army Commander in the rank of Lieutenant General. He had done extremely well during the last few days, but had twisted his ankle in one of the pleasure skiing holidays. His instructor Lieutenant Dhillon told him that he had attained more than the required proficiency for the basic course and instead of reporting to the hospital, he should come and stand on the slope with the score so that he could complete the course.

One day, the senior instructor Captain Hoon, who indeed was a bully came to their squad and saw him not skiing. He explained to him that his instructor had directed him to go down the slope. He had no other choice but to go down and damaged his twisted ankle. However, his instructor Lieutenant Dhillon told the Commandant, then Major Gurjeet Singh that Bull Kumar had obtained the required proficiency in skiing. Thus, he not only got him the certificate for basic course, but he also highly recommended him for the Advance Course.

Luckily, Bull Kumar availed 2 months leave and during that period his ankle healed. The next winter when he was selected for the Advance Course, his CO decided not to let him go for the course as he thought that skiing was professionally not useful for his career.

Luckily, snow came a little early that year and when the Division Commander General K S Thimayya, DSO. who was from the Kumaon

Regiment came to visit them, he thought of a plan that worked very well. He gave a rupee to the unit carpenter and asked him to build skis for him. To help the direction of slopes he made a groove in the centre of the skis, and then he put 2 big blocks on the ski heads and made a curve so that the wooden plank won't get stuck on small bumps. For his ski bindings, Bull Kumar put nails on the side of the skis and tied that to his boots with the help of line bedding. He got some bamboos to improvise the ski sticks and then waited for the right occasion.

The moment he saw General Thimayya and his CO coming up the snow covered road where they could see him, he started skiing down. He fell down once or twice but got up and reached the spot where they were standing. General Thimayya was really impressed by the contraption of the skis that he had improvised and told his CO to send him on the Advance Course.

In the winter of 1957-58, he was happily enjoying his skiing in Gulmarg and they were having great fun. It was near the end of the course that General Thimayya invited an Italian skier to visit Winter Warfare School and though he was over 50 years, he used to climb up to Alpather and come down swinging like a dancer. They all thought that it was because of his good skis that he could ski so well. So, all the officers started throwing parties for him hoping that he would leave his skis with them when he left. At the end of the course, he gave a big lavish party for all the officers but took his skis with him. They were all disappointed as they only had hickory skis made by the Indian Ordnance Factory that had fixed spring bindings unlike the one used by the Italian which had automatic bindings that would come off with little pressure and not hurt the user's limbs.

After the course was over, while they were trying to race the Italian instructor, Bull Kumar had broken his ankle again. He went to the hospital and his ankle was plastered for 6 weeks. After that he got two months medical leave and in all he was out of the Battalion for about 4 months. He really dreaded going back to the unit after such a long absence. Luckily, his posting order came to the Kumaon Regimental Center (KRC) from where he did his Platoon Weapons Course. As he had done very well in the Weapons Course, he thought that he would be appointed as the Weapons Training Officer (WTO), but instead he was made the Physical Training Officer (PTO) by the Commandant.

In 1958 while he was at the KRC, a mountaineering course at HMI, Darjeeling was offered to the Centre. At that time nobody knew anything about mountaineering and when he opted for the course, he was refused. By then he had learnt the trick and he asked his Commandant for an interview and told him that he would forgo his annual leave to do the course. The Commandant replied, in that case he had no objection in his attending the course and thus, he landed up in the HMI, Darjeeling to do the course. At that time the Principal of HMI was on a Cho-Oyu expedition in Nepal and Tenzing Norgay looked after the course as he was the Director of Field Training. Infact, HMI was started by Pandit Jawaharlal Nehru after Tenzing Norgay climbed Mount Everest in 1953. Tenzing remained as the Director of Field Training and was given extension of service even after his retirement.

BULL KUMAR'S FIRST EXPEDITION – TRISHUL

Tenzing Norgay was not only a great climber but was also a very close friend of Bull Kumar. Though Bull Kumar was only marginally senior to his course mates, Tenzing called him one day and told him that he would be in charge of the course for the officers. He also advised Bull Kumar to maintain some distance from his course mates to ensure discipline.

Tenzing suggested an expedition to Trishul to Bull Kumar. Every evening Tenzing used to call him to his tent and they had briefings and discussions on the imminent expedition. Bull Kumar started thinking about organizing the expedition and they discussed equipment, training, funding, and associated matters. He also got another YO Lieutenant YK Yadav, from artillery involved. The third member of the team was Sub Lieutenant PP Mehta from the Indian Navy. Most evenings were spent planning the expedition to Trishul. That mountain was pretty high in altitude – 23,360 feet - climbed for the first time by Youngstaff in 1907, and was promising to be a daunting task for the Indian heroes!

They returned to Darjeeling after the training and were enjoying the city after twenty eight days in the wilderness. Major Nandu Jayal, the first Principal of the HMI, Darjeeling had died on Cho-Oyu expedition due to pulmonary oedema and Colonel (later Brigadier) Gyan Singh who was later the leader of the First Indian Everest Expedition in 1960 had taken over. During the passing out address by Tenzing Norgay, he presented Bull Kumar with a pair of socks that his wife had knitted from the fur of Apso Dogs. He had about 35 dogs with him.

As they were discussing funding of the Trishul expedition, one evening Tenzing Norgay invited Bull Kumar to his house as he was expecting a correspondent of the New York Times. It was five in the evening and Tenzing Norgay showed Bull Kumar his cupboard where a large collection of whiskey, various kinds of liquor, champagne, which were all presented to him by various visitors were stocked. He invited Bull Kumar and the correspondent to enjoy drinks. According to the Indian Army custom, it was still early to start drinking. So, Bull Kumar was unable to offer beer, gin and vodka to the correspondent as these were the pre-lunch drinks while the whiskey was the pre-dinner drink. So, they settled down with champagne and enjoyed the evening.

During the meeting, Bull Kumar brought up the subject of the New York Times sponsoring the Trishul expedition and the correspondent asked how much the whole expedition would cost. At that time the estimated expenditure was about Rs 15,000/-. As the New York Times was a very reputed newspaper, Bull Kumar decided to raise it a bit and said the total expenditure would be about Rs 20,000/-.

The correspondent quickly converted the amount into dollars and told them that if he gave that budget to the New York Times, they would not think that the expedition was an adequately big venture. So, he raised the amount to Rs. 50,000/- and whatever amount was leftover could be donated to the Sherpa Welfare Association that Tenzing Norgay had wisely started for educating the Sherpa children.

Bull Kumar was extremely satisfied so he conveyed the good news to Colonel Gyan Singh, the newly appointed Principal of the HMI, Darjeeling. The Principal was jubilant at the prospects of the HMI students going on an independent expedition! He spoke with Brigadier (later Lieutenant General) Bhagat Singh, VC – then the Director of Military Intelligence.

Brigadier Bhagat, VC gave an earful to Colonel Gyan Singh by telling him that within a few months of taking over a civilian job, he had forgotten all the rules and regulations of the Army, whereby every officer on commissioning signed a certificate stating firstly, that he would not steal affections of another brother officer's wife, and secondly, he would not have any contact with correspondents and foreign nationals. He continued by ranting how could he take an expedition by using money from a foreign correspondent? So, every thing was back to square one.

However, Colonel Gyan Singh was a very determined officer and he did not want the expedition to fall through. He telephoned General KS Thimayya, the COAS, under whom he had served as Officer Incharge of logistics when he was General Officer Commanding (GOC) 19 Infantry Division. This division had saved Ladakh by sending armoured vehicles across the Zojila Pass to contain 'Jihadis' supported by the Pakistani Army to annex Jammu & Kashmir. It was technically a non conventional operational move and the Jihadis who were not equipped to fight against tanks withdrew.

General Thimayya, one of the very famous and great Generals India has produced, was an exceptionally generous senior officer too. He asked the respective centre commandants of the expedition members to donate Rs. 2,000/- each and also asked them to give the officers necessary leave.

Bull Kumar's Centre Commandant Lieutenant Colonel Ram Singh from Rajasthan had objected to his doing the Mountaineering course as there was no professional gain in doing that course. But, Bull got his permission because he had surrendered his annual leave. And there, he was asked, to not only give Bull Kumar his annual leaves but also pay Rs. 2000/- from the regimental fund. Bull Kumar became a very unpopular youngster in the Kumaon Regimental Centre (KRC).

They had 3 members of the expedition and one of them was from the Indian Navy. The Chief of the Naval Staff (CONS) had also followed the example of the Chief of Army Staff (COAS) and gave Rs. 2,000/- for Sub Lieutenant P. P. Mehta. However, the amount donated was not enough to defray the entire expenditure and the members of the expedition had to contribute some amount personally.

Bull Kumar remembers having to sell his motorbike to raise the funds and also selling his transistor radio. Later he had to borrow one for the expedition to get the weather forecasts. Finally his first expedition was on the way. In those days they had to start marching from Chamoli, a stage before Joshimath, and then go on the mules up till Lata village from where porters from villages of Reni and Lata were hired. One of the main obstacles in organizing an expedition in those days used to be arranging the porters.

Shri Gurdial Singh of Doon School, Dehradun, who was the first Indian to climb Trishul, was also leading an expedition to Nanda Devi that year. Bull Kumar asked the porter Sirdar whether their expedition would clash with that of Gurdial Singh. He told Bull Kumar that it wouldn't if they left their loads till Deodi, halfway to Nanda Devi sanctuary. So, they decided to return most of the porters required by Gurdial Singh's expedition at Deodi and carried on with the remaining porters who would do double shift to Deodi and Base Camp. However, later Bull Kumar was blamed by Gurdial Singh for stealing his porters.

The first day's climb from Lata to Kharak was about 4,000 feet in elevation and being a very steep gradient, was the most exhausting. After that they crossed the Dharansi Pass (13,940 feet) which also marked the outer sanctuary of Nanda Devi. From there, they descended down to Dhibru Ghetta where there was a very fascinating pass in the middle of the jungle. From there one descended down and spent sometime in laying an improvised log bridge over Rishi Ganga; this was real life true blue adventure, serious adventure; crossed over to Deodi from where they returned the porters required by Gurdial Singh's expedition; and carried on with their expedition with whatever porters were left.

They went one more stage before getting to their Base Camp in the Trishul Nala (stream). From Base Camp climbers returned all the porters except three high altitude porters who helped them to carry their loads to the next, higher camp.

One of the most difficult parts in planning an expedition in those days was the arrangement of mountaineering equipment. Bull remembered to have his windproof mittens up and had to go to the cobbler who took his hand impression and stitched a pair of mittens from calf leather. They never had proper mountaineering boots so they used regular ammunition boots issued by the Army. To keep their feet warm, they had put on cotton socks, a polythene bag over them; followed by thick woolen socks and another polythene bag so that the main insulation did not get wet. They also got crampons made by a blacksmith, though they did not use them much as Trishul (23,360 feet) was not an icy mountain. For climbing the summit of Trishul, they just had to slog up making new steps in the soft snow and they were lucky that they could climb the summit that way. As Bull was an expert in skiing, he felt Trishul provided an excellent slope to ski on.

On their return journey, they went to Nainital, a very popular hill station in northern India where on the Lake Club they made friends with some young ladies. One of lady's fathers was a correspondent of the Times of India and he took their pictures with their beards. Before they reached Ranikhet, their picture was splashed on the front page of the Times of India. That changed the entire atmosphere in the KRC with the others championing the bravery and adventure of the crew and a hearty bonhomie and vicarious feelings of victory doing the rounds; and consequently they were treated as heroes on their return.

Young cadet Kumar of IMA and young Mridula before their marriage

Col. Kumar's parents and younger brother Kiran Inder Kumar

The three brothers and sister (Left to right: Maj. Davinder Kumar, Col. Narinder Kumar, Saroj Passey, Brig. KI Kumar)

The family (Left to right: Maj. Gen. Sundan, Brig. KI Kumar, Mrs. K I Kumar, Col. N. Kumar, Saroj Passey, Mrs. Mridula Kumar, Niece Vaani Kapoor, Mrs. Toshi Sudan, Sohan Passey – Brother-in-law)

Col & Mrs. Kumar after their marriage

Yash Karan Singh and Shailaja, Col Kumar's son-in-law and daughter

Col. Kumar's granddaughter Sia (Named after Siachen)

Col. Kumar's son Akshay Kumar who was the first to descent down Brahmaputra from Tuiting to Pashi Ghat in Assam with local girls from Arunachal

Dilshad, Saira and Akshay Kumar – Col Kumar's daughter-in-law, grand daughter and son

Bamboo and ghee tin raft, with which Col. Kumar rafted down Teesta

Mrs. Indira Gandhi watches the rock climbing demonstration organised by Col. Kumar, Principal of HMI Darjeeling

Col. Kumar, the Principal of HMI Darjeeling with Tenzing Norgay, Director of Field Training

Members of the 1958 Trisul Expedition (Left to right: Lt Y K Yadav, Lt N Kumar, Sub Lt of Indian Navy PP Mehta who got frostbite during the expedition

Trishul Peak

Dakota plane crash at Barahoti which was bringing rations for Col Kumar and team

Capt. Kumar was the first to be awarded AVSM for his successful expedition to Barahoti

EVEREST - 1960

Meanwhile, some highly influential and respected senior bureaucrats in the Government of India such as Shri SS Khera – former Cabinet Secretary, Shri HC Sarin – Ex Defence Secretary and senior Army officers like General Williams of the Engineers, decided to send an Indian expedition to Mount Everest in 1960. All the mountaineers got a letter asking if they were willing to go on the expedition and how much money they would be able to pay towards the cost of the expedition.

Kumar was then posted as Brigade Transport Officer (BTO) in Lorried Infantry Brigade at Mathura. One day, in the Officers' Mess, he was sitting alone in the bar having a drink when Brigadier KC Khanna, Brigade Commander, entered the bar. He saw Bull and asked, 'Young boy, you seemed to be worried about something'.

Then, Bull told him that he had just led a successful expedition to Trishul and now he had been asked to be a member of the Everest expedition and also how much money would he be able to contribute towards the cost for the expedition. He also apprised his Brigade Commander that as he was still paying bills for his Trishul expedition and how he was unable to raise any money for the Everest expedition. Brigade Commander asked him to see him in his office the next day in the morning.

The next morning the Brigade Commander had consulted his staff and had rung up Major General Badhwar, General Officer Commanding (GOC) 1 Armoured Division located at Jhansi. They had both decided to contribute Rs 20,000/- each and his Battalion Commander was asked to add Rs. 10,000/- from the Regimental funds. Thus, when Kumar met him, his Brigade Commander told him to inform the organizers of the Everest expedition that he would contribute a handsome amount of Rs. 50,000/-. Later on Bull Kumar realized that he was the only member who had volunteered to contribute towards the expedition.

In 1959, all the mountaineers who were selected had to take part in a pre-Everest expedition for the final selection of the team. That pre-expedition was held in western Sikkim in the area of Kabru Dome. All the members got together at the HMI, Darjeeling where the accommodation consisted of 3 double decker bunk beds to a room. It so happened that in Bull Kumar's room there were 5 boys from Punjab and one from Darjeeling whose name was Suba. In those days an agitation was going on in Punjab for a Punjabi Suba (Punjabi Province). So their room was nicknamed 'Punjabi Suba'.

As Bull Kumar was keen to go to Everest even as an Administrative Officer, he took upon himself all responsibilities of handling food logistics, equipment, preparing load manifestos and distributing the same to the porters, to impress Brigadier Gyan Singh who was nominated as the leader of the expedition.

There was stiff competition for the selection of the expedition team. The members included India's boxing champions, instructors from physical training schools and other sports personalities. However, as the pre-expedition progressed and aspirants hit high altitudes, many of them fell out due to high altitude sickness. Tenzing Norgay who was the first Indian to climb the Everest along with Hillary was incharge of the training camp.

A rather special incident that Bull Kumar still remembers was that, when one group was leaving for Camp-I from the Base Camp, Tenzing decided to cancel departure. Surprisingly a gigantic avalanche had come galloping down upon the route that was to be followed by the group on the same day. Though Tenzing had no scientific experience to forecast avalanches; he could sense them; was it clairvoyance or some extra sensory perception? Whatever the explanation it certainly saved the day!

Bull Kumar did rather well on the mountain and was quite certain that he would be part of the expedition team. So, for the return journey he handed over his administrative duties to other members who had not done so well in the pre-expedition, so that they had a fair chance to be selected as Administrative Officers for the expedition.

After the team returned from the mountain to the HMI, Darjeeling, selection for the Everest expedition team was made. Apart from three Sherpa instructors from the HMI, Bull Kumar topped the list for the expedition team. These three Sherpa instructors were among the seven Sherpas who had carried loads to the last camp in the First successful Everest expedition in 1953 with Col John Hunt as the leader.

They were also given duties for preparations for the 1960 expeditions and Bull Kumar was assigned the task of arranging equipment including oxygen. In those days, there was a tremendous shortage of mountaineering equipment in India and the Government took that opportunity to manufacture most of it in the country and import just seven sets of mountaineering equipment for extremely high altitude work.

As the Joint Secretary (Defence) was involved in the expedition, Bull Kumar had direct approach to all Ordnance Factories and he used to visit some to give samples of the mountaineering equipment needed, get back the prototypes for testing and go back to those factories to give them the feedback and carry out modifications if any needed. The one raw material that they had to import was Eider Down (Goose feathers) used for stitching parkas, gloves, sleeping bags, etc.

After great effort, the Paragon Industries could make the cloth that could hold feathers and in turn they gave the cloth to Ordnance Factory, Shahjahanpur. Their rucksacks were manufactured by the Ordnance Factory, Kanpur and iron mongery by the Ordnance Factory, Jabalpur. Bata was very kind to produce trekking and climbing boots for the climbers free of cost. But, after the trial report they had found that the Indian leather was not good enough for mountaineering boots and they had to be imported. However, it made the beginning of the manufacturing of mountaineering equipment in India. That experience came in very handy for the Ordnance Factories which had to manufacture equipment for troops after the 1962 War with China.

Lieutenant M.S Kohli was handling the victuals for the expedition, while Captain S.C Nanda was the Signal Officer elected to handle communication. By February 1960, the well trained expedition team, proudly equipped with Indian equipment was ready to leave for the 1st ever Indian Everest Expedition. While trekking, apart from handling equipment Kumar was asked to manage porters and Sherpas too.

In those days, there were very few camping sites at various stages to march up to Base Camp from the border town of Jaynagar. There was another small foreign expedition going along with them. On the very first day of the marching for the expedition, they lost the way and had to do some extra miles to get to the correct stage. Though it was a costly mistake, it had a silver lining; they got an extra day for acclimatization.

Once the small expedition that was going along with them occupied the main camping ground and they had to scatter their tents all around. It was a good lesson to learn as their expedition was a very big one with five climbing members, four administrative members like the doctor, signal officer etc along with 50 Sherpas and 700 porters. So, for forthcoming marches, they used to send an advance party ahead of the main expedition to occupy main camping grounds. However, they did not deprive the other expeditions from camping on the periphery of the camping ground.

It is said that 'Shunya' ie. Numerical '0' was invented in India. Bull Kumar made great use of that invention and instead of numbering his first load as '1', he made it '0' so that once the porter had to lay down the loads after the march for counting, the first lot would be up to 9, the second one up to 19, third line up to 29 and so on so that porters could easily identify the first numerical and put their load there. The Sirdars used to put them in order and it was very easy to count. With this counting system, they carried all the loads to Base Camp without any loss. The loads that were used before the Base Camp were marked in different colours and thus the loads required for climbing reached the Base Camp without any problem.

Enroute, many a time, members found some menfolk, got drunk before reaching their destinations, as they passed through villages; and the Sherpa ladies had to go back to help them carry their load to Base Camp. The local drink 'chhang' and the other brewed from millet was equivalent of beer. In the higher altitude, Sherpas call it 'Thumpa' and was drunk using a straw from a wooden jar and the host kept on pouring water into it as one sat down sipping bit by bit. The drinking session would last up to an hour or an hour and a half. A person did not feel the intoxicating effect of the local brewat once, but when one stood up and hit the cold, he would start swaying.

They had quite an experience with the Sherpa porters. Once, one of the lady porters carrying loads weighing 25 kgs excused herself from the caravan line and went along the bushes with some other ladies and she came back with a baby that she had delivered in the bushes. Though she was prepared to go on carrying the loads as she did not want to lose her wages, the members redistributed her loads to others and she kept marching along the caravan till she reached her village in Namche Bazaar. Such is the hard life that Sherpas live!

On about the 15th day of the approach march they were having a lunch break on the banks of river Kosi. They removed their rucksacks and joined General Thondup - their great cook. After his lunch Brigadier Gyan Singh went back to the place where Bull Kumar's and his rucksacks were resting and by mistake he picked up Bull Kumar's rucksack. He could not even move Bull's rucksack and walked away and stood near a tree. Finishing his lunch, Bull Kumar came up and picked up his rucksack and after setting it on his shoulders, he started walking normally.

After about half a km, Brigadier Gyan Singh caught up with him and asked him, 'how heavy is your rucksack?' Bull Kumar was surprised at the question; and replied '75 pounds Sir'. Starting from normal 30 pounds, he had been adding 4-5 pounds every day according to the length of the track. The Brigadier was obviously impressed and asked him to go slow and not burn himself out before reaching the summit. Bull Kumar took his advice seriously and next five days he did not add any extra load to his rucksack.

Those were the days of foreign exchange shortages and they were allowed to import only eight sets of clothing for higher altitude and for the additional requirements, they had to depend upon Indian equipment. They had selected red colour for most of the equipment for better visibility. Bull Kumar was one of the potential summiteers, and the summiteers were jokingly called "*Lal langotias*" by team members.

At the Base Camp all the porters left their loads and went back and the team members were left with just 50 Sherpas. After establishment of the Base Camp, the work on the ice wall started. Captain MS Kohli and Sherpa Ang Tomba were the first two to enter the deadly ice wall and returned to Base Camp after establishing Camp-I in the middle of the ice wall at about 19,000 ft. Namgyal and Bull Kumar replaced Captain Kohli at Camp-I and they started working towards Camp-II. Bridging crevasses, fixing ropes, cutting huge blocks of ice mushrooms, they opened the route to Camp- II at 20,000 ft.

They were instructed to return after establishing Camp-II, but when they saw the *cwm* (highly perched up valley in Welsh language), It looked flat and simple and they decided to go up to Camp-III (Advance Base Camp) covering that distance before breakfast. The weather was good and time was with them. Bull Kumar was toying with the idea of moving to Camp-IV. He looked at Namgyal inquiringly; who gave him an approving smile. Namgyal was one of the greatest human beings Bull Kumar had ever met - tough, polite, dependable and self sacrificing.

They carried on and by one o'clock they had reached the site of Camp below Lhotse Face. They dumped some stuff and started their return journey. Till then they were physically very fit. But by coming down, the glacier latitude, the direct hot sun, dehydration and effects of high altitude had started taking their toll. Their water had finished and much against medical advice they were forced to eat and drink snow. Bull Kumar hoped that staying at Camp-I would make both of them better but they had gone too high too quickly - 5,000 ft in 2 days without acclimatization at each stage and spending another night in poor physical conditions.

Bull Kumar got up with a very severe headache and nausea in the morning. Namgyal made tea for both of them, but even drinking tea was an effort in his condition. Namgyal, who had even carried loads to the last camp (27,500 ft) in 1953 Everest Expedition with Lord Hunt so that Hillary and Tenzing could get to the summit, was also feeling the ill effects of high altitude, though not as severely as Bull Kumar.

The downward journey through the treacherous ice wall took 14 long hours which was normally covered in 2 hours by fit climbers. Every few yards they would sit down, as if all their energy has been sapped out of their systems and the descent became a torture and many a time Bull Kumar prayed that God will do something like opening a big crevasse to swallow them or send an avalanche to bury them because he wanted to end his life.

With the passage of time, their stomach became empty of fluids and they started to vomit yellow bile. The pain was not only in their heads and throats but even in their cramping, contracting stomachs. Coming down step by step, they crossed Camp - I at 3.00 pm They did not have a wireless set, which would have been so heavy in those days and they would have rather carried an extra tent.

Brigadier Gyan Singh got extremely worried and sent up some help with torches and tea, it was a miracle that they got down to the Base Camp. The whole night Bull Kumar was kept on fluids by Captain Das and he was sent down to Phereche at about 15,000 ft for recovery. Captain Das accompanied him. The expedition had lost one of its summiteers as Bull Kumar was totally written off by the team as they thought that he would be sent down to Delhi and by the time he recovered from his illness the expedition would be over.

For the whole week the team had good weather. Work on the mountain continued and route to the South Col was opened. While opening the route they had to put up only one more camp (Camp-V) before they reached the South Col. The South Col is windy sharp-edged pass between Mount Everest and Lhoste where climbers attempting to climb Everest from the southeast ridge in Nepal usually set their final camp.

Bull Kumar Kumar was getting all the news at Phareche cursing himself for his folly and bad luck. He had worked very hard for that expedition and had also got all the special equipment manufactured from the Ordnance Factories, getting oxygen cylinders and various adapters made. The dome tent used by the Army in later days was also designed by him for the expedition.

Mercifully his recovery was quick. One day, he received a call that the leader Brigadier Gyan Singh had already left Base Camp to go to Advance Base Camp, where he would announce the summit parties. Captain Das would not let him go for the expedition again after what he had suffered. But, he was able to convince him. By the time he reached the Base Camp, Brigadier Gyan Singh had already reached Camp-II. The next day he was to reach the Advance Base Camp and announce the summit parties.

Bull Kumar got hold of couple of Sherpas and dashed up the mountain and was surprised that he was covering good ground. At about 2 pm having reached Camps I & II, he overtook Brigadier Gyan Singh's party and reached the Advance Base Camp. He found himself very fit. He got some tea made and came out to receive Gyan Singh's party. Nobody noticed anything until suddenly they realized that it was Bull Kumar who had received them. When did he overtake them, they wondered! Some members thought for a moment that that it was his ghost. Brigadier Gyan Singh was pleasantly surprised too! That evening saw a lot of discussion in the Mess tent. Some of Bull Kumar's friends argued that he was fit enough to become a summiteer. They all agreed that he was one of the toughest of the lot but doubted his capacity to take to the high altitude that had already rejected him.

The next day Gyan sent Kohli, Bull Kumar and Nima Sherpa up with an excuse to check the old oxygen bottles left at South Col. His idea clearly was to see that how Kohli and Bull Kumar took altitude. They reached Camp-V and after having had their share of food and having prepared tea they awaited the Sherpa party who were supposed to return from South Col. But the Sherpa party did not come back. Bull Kumar informed Brigadier Gyan Singh who was at Advance Base Camp and he was worried about them. Brigadier asked him what he could do. Bull Kumar told him that he was fit enough to go and search for them.

Leaving Kohli at the Camp-V, he slowly climbed up along the fixed rope. After about half an hour he heard some noises just across him in a crevasse where Phu Dorjee and 8 Sherpas had taken shelter. Nima went up and gave some tea and they asked them not to come out and they returned to Camp-V. Brigadier Gyan Singh was very happy to get the news that the party was safe and sound. The doubts about their being able to stand high altitude were put to rest. Brigadier's doubts about Bull's body to withstand the rigours of the high altitude had also vanished. They were told that there was no need to go to South Col anymore and bring down the Sherpa Party as some of them were suffering from frostbite.

Summit parties were announced and Bull Kumar was delighted to get his place in the 1st Summit party along with Nawang Gombu and Sonam Gyatso. Captain Kohli and Ang Tamba were selected for the second Summit party. On the 24th of May, the first summit party moved to the last camp at approximately 28,000 ft. On the 25th of May, they wanted to start at 5.00 am but the wind was so strong that they had to delay their departure by a couple of hours. They all felt very fit despite having had very little sleep at night. After half an hour of climbing, Nawang Gombu, who was leading the rope stood motionless stooping heavily on his ice axe. Bull Kumar caught up with him and was aghast at the sight of his oxygen bladder blown out like a balloon. Bull Kumar instantly knew what had happened. The intake valve had frozen in the strong winds. He quickly removed his gas mask and with a Swiss army knife removed the frozen ice on the valve. The balloon immediately subsided and Nawang Gombu started climbing normally.

The wind was so strong that it was very difficult to hold on the surface of the ground and not be blown out in the atmosphere. In the process the middle finger of Bull Kumar right hand got a nip of the frost bite. But that did not bother him at all. He looked at the time; it was 11:15 am and they had hardly covered 300-400 ft. The strong winds made the progress very slow. Most of the time they were climbing with their heads turned towards east to avoid the strong westerlies. Then all of a sudden, Nawang Gombu halted. Sonam Gyatso caught up with him and Bull Kumar was just a few feet below them. Perhaps the most important decision of the expedition and their lives was taken without saying a word and slowly they started moving down. Just few hundred feet below the last camp Bull Kumar slipped on an icy patch but Sonam Gyatso was very alert and held him tightly. 25 years later Bull Kumar's younger brother Kiran Kumar while trying to climb Mount Everest solo, slipped at that very point and in few seconds his shattered body had been found at Advance Base Camp, almost a drop of 7000 vertical feet.

The weather remained so bad that Kohli and Ang Tamba could not move up from South Col. The next day Kumar came down from South Col to Base Camp and all the time he was wondering, how to face the team and the Sherpas who had worked so hard to put him at the last camp for the summit attempt. They were all sitting in the mess tent and people were trying to console him when AJS Grewal, popularly known as Khalifa, addressed him and told Bull Kumar not to worry. Brigadier Gyan Singh was so efficient that he already had two messages ready to send to IMF, one if he had succeeded and the other if he had failed. Everybody had a hearty laugh on that stress buster and the scene became more bearable in the mess tent.

The King of Nepal was not very happy as at that time there was a dispute going on between China and Nepal about the ownership of Mount Everest. However, the dispute was settled between the two countries by an agreement that Everest belonged to both the countries - the northern side to China and the southern side to Nepal. Well, many people have asked Bull Kumar if he was disappointed for having missed the summit. He always musingly replied, "Disappointed, yes but not dissatisfied" as they had done their best. And once one has given one's best, one can not do any better.

Before leaving Kathmandu, they booked the mountain for 1962 and returned to Delhi. Bull went back to his unit located in Tangdhar across the Nasthachun Pass where his Battalion was holding a very high peak on the cease fire line.

BARAHOTI - 1961

After his Everest attempt, two things happened. Firstly, his CO took him wherever there was a party in order to show off that he was one of the Everest climbers. Secondly, whenever any officer had to go for payment to troops deployed on higher pickets or if there was any court of enquiry to be held at any picket, Bull Kumar was detailed for such duties. His CO thought that Bull Kumar must climb 10 times more than he climbed for Mount Everest, but of course at much lower altitude and that was how Bull Kumar could build up his climbing muscles.

The IMF planned 3 expeditions to select a team for the 1962 Everest expedition and Bull Kumar's share came to a very difficult mountain - Nilkantha. It was not very high, but technically it was very different. The two other expeditions were to Panch Chuli in the Kumaon Himalayas and the other to Annapurna-3 in Nepal. The three leaders selected the teams for their respective expeditions in consultation with the sponsoring committee of the IMF.

While Kumar was preparing for the Nilkantha expedition, one evening in the Officers' Mess when everybody had couple of drinks, the Adjutant of the school came to him and said, "Young man, get ready to go to Army Headquarters in Delhi tomorrow". Since he had played that prank himself many times on others, he continued to order another round of drinks and the party carried on.

After an hour, the Adjutant came back to Bull Kumar with the same message and he felt that another round of drinks was due from him. However, after the dinner the Commandant Maj (later Brigadier) Gurjeet Singh came to him and said, "Come boy start packing" and seeing that he was dead serious even after the grand party Bull Kumar had thrown, he was convinced that it was not a prank at all.

His immediate reaction, however, was not one of excitement or anticipation at what lay ahead, but of "fear". Why had the wrath of the Army Headquarters landed on a junior Captain? He just could not help but think of all the misdeeds that he had been a party to, but surely none of them was serious enough to warrant action directly from the Army Headquarters.

It was in that confused state of mind that he reached Srinagar at 8 in the morning. The ADC to Major General AS Pathania, the then Division Commander, was waiting to meet him. After a frugal breakfast (he was so nervous that he could hardly eat anything), the ADC escorted him to the Operations Room. It had a huge sand model of the area and several maps marked with red, blue and green pins on display.

Just when he was about to reach the end of his patience, the General addressed him in his unmistakably hoarse voice, "Young man, you have been selected by the Army Headquarters to carry out a task of national importance. Do not hesitate to ask me for any help that you may need. I will feel extremely proud if you can carry out the mission".

Though these words were heartening they did not solve the mystery as to his purpose of going to the Army Headquarters. Many times in his short career in the Army, he was told that YO's were not to be heard, only to be seen. In other words, it meant to keep one's mouth shut in the presence of senior officers. Even though that lesson had been well learnt, he could not restrain himself from asking, "What is the mission, Sir?" The General gave him a hard look, obviously surprised at his question and answered, "So far everything is a secret. I too do not know what the mission is about".

Bull Kumar was informed by the General that a special Dakota was being flown in from Delhi to pick him up from Srinagar. Out of all vagueness surrounding him, that was the first heartening piece of news.

"Wow, a chartered aeroplane flight just for me?" he thought with a small smile playing on his lips. However, his feelings of joy were short lived. The weather over Banihal pass at 8,985 feet was bad and the Dakota for him could not cross over the Pir Panjal range to get him out of the Kashmir Valley.

Since his "mission" could not wait for the weather to clear he was assigned a small convoy of jeeps. While he was bundled into one, a spare jeep followed him in case of mechanical failure. That spare jeep was followed by another one with a mechanic. Nobody was taking any chances about his reaching his destination, the rail head at Pathankot on time.

No sooner had he got on to the jeep that he realized that the timings of the mission were significant. The driver was told to forget about Army speed limits and drive as fast as the meandering roads would allow safely. He was given instructions not to stop anywhere on the way. They reached Udhampur at around 3:00 pm and were forced to halt by the military police. A Major alighted from the jeep and ordered him to follow him to the Corps Headquarters. He tried to impress upon him that he had no time to waste, but without effect. Bull Kumar was told that the senior General wanted to see him and there was no way he could say no to that.

What happened there was a repetition of the scene that was enacted in General Pathania's room in Srinagar. However, even Lieutenant General Verma, the then Corps Commander had no inkling of the task that lay ahead of him. The only thing he knew was that the task was of "national importance" and he was proud of the fact that an Army Officer of his formation had been selected for the job.

The moment he learnt that the General knew nothing about his mission, Bull Kumar lost interest in the interview and kept stealing glances at the wall clock as the departure time of the train he was to take to Delhi was inching closer. The General perhaps sensed what was going on in Bull's mind and he told Bull that the Srinagar Express would not leave Pathankot without him which was something he could not digest - a train was waiting for him, an insignificant Captain of the Indian Army.

When he reached Pathankot railway station, he heard a heated discussion going on between the Station Master and the RTO (the Army representative for the control of military traffic). The train was already two hours late and the Station Master did not want any more delay. However, Bull Kumar decided to make the most of his exalted position and demanded a cup of tea before he boarded the train. He was, however, not being totally selfish as he badly needed a cup of tea because he had neither eaten nor had anything to drink since his meagre breakfast early in the morning. Despite the protests of the Station Master, he gulped down his tea and finally boarded the rather delayed train.

With nothing to occupy his mind on the journey, his thoughts started moving from one thing to another. He was gradually convinced that his selection was based on his mountaineering experience. He was the first non Sherpa Indian to have reached the altitude of 28,300 ft, just 700 ft short of the Summit of Everest last year. He could then begin to guess where the task would take him. He did not know its purpose and what benefit it would have for the country. The only thing that seemed certain to him at that point was that it would lead him to the mountains and to difficult ones at that.

Not that Bull Kumar had any problems with mountains. In fact, he was already beginning to get some pleasure from where the mission was going to take him. He had once been branded as a non-serious officer by his CO when he had volunteered for Basic Mountaineering Course at the HMI.

The intervening years had taken him to various mountains. In the regiment, when he had returned from his Everest Expedition in 1960, he had lot of unsolicited advice from his friends: "It was a high time he got down to some serious soldiering and stopped poodle faking".

They, of course, meant well but their words had hurt Bull Kumar because he had done no mountaineering at the cost of his military duties and that his mountaineering speciality - the so called poodle faking was going to come in handy for the Army. He flushed with pride as the train chugged towards Delhi. He, with only six years of service, had been chosen over the heads of some of the most capable officers in the Indian Army for an extremely special and gritty task.

He reached Delhi in the early hours of the morning. He tried hard to be on time, but he reached the Military Operations Room at the Army Headquarters fifteen minutes late. In those 15 minutes, he was later informed, Gulmarg, Udhampur and Srinagar had been contacted on the flash call systems to find out his whereabouts. Somehow, no one had thought about contacting the Queen Victoria Officers' Mess - the normal habitat of single officers staying in Delhi.

When Bull Kumar reached the Army Headquarters, he saw the Director of Military Operations pacing up and down in his room. As soon as he saw him, he grabbed him by the arm and pushed him into the room where lots of senior officers were assembled. He still remembers late Lieutenant General BM Kaul, the then Chief of General Staff, late Mr HC Sarin, ICS, Joint Secretary in the Ministry of Defence, late Mr PK Dave - Joint Secretary, Ministry of External Affairs who later become the Lieutenant Governor of Delhi waiting for him anxiously.

General Kaul introduced him to the others with the words, "Gentleman that is Captain Kumar. He is the first Indian to have reached the dizzy heights of 28,300 ft on the Everest. If he can not deliver the goods, no one else can".

He felt good at the introduction and flashed a silly smile. To cope with such situations was something beyond him.

He would have felt a lot better if he had known even a few details about the mission he was soon to embark on. But he was confident of one thing; the journey of his life was soon heading to reach its first definitive destination.

Dave gave all those assembled, a background to the dispute between the two major Asian Countries- India and China wherein the disputed area in that case, lay somewhere in the Central Himalayas. The area was called **Barahoti** by the Indians, and **Wuje** by the Chinese. The talk was too full of International legalities for it to register completely with Bull Kumar. However, finally Mr. Dave mentioned the part that was of interest to him - Barahoti lay well inside the Indian Territory.

But every year after the snow melted, the Chinese came and occupied that area. On the Chinese side there were no difficult passes to cross to reach Barahoti while on the Indian side, the terrain was far more difficult. Even a knowledgeable and capable person like Brigadier Gyan Singh, leader of the First Indian Everest Expedition, felt that it would be difficult, if not impossible, to cross the great Himalayan range from the Indian side before June-July. The main obstacle to Barahoti was the 18,000 ft high Chorhoti Pass.

The Chinese had been occupying not only Barahoti but had been establishing their posts all along the Indian border and claiming the area as their own. At that time since there was no war between the two countries, it was a claim to a territory that was rightfully ours. Indian leaders had decided to try out what was called by some as the "Forward Policy". They wanted to see the Chinese reaction if India tried to establish its posts in the disputed areas. That was precisely what Bull Kumar's mission was going to be; to establish posts in Barahoti.

Crossing Chorhoti was the first step to reaching Barahoti. He was to lead a party of 50 men across the pass while it was still under heavy snow. He was to go very deep into the mountains, there was no way any reinforcements could reach him. Administratively, the venture had only a 10 percent chance of success.

For the success of any mountaineering expedition a pyramid is planned. The base is big and wide while the top is just a point. During the British Everest Expedition of nearly 800 people with 50 Sherpas, 15 members had been able to put only two climbers on the summit of Everest.

Barahoti, however, called for an entirely different planning. Here Kumar had to take the entire Base Camp with him. And he had only 15 to 25 days in which to plan the entire operation.

While all that was being discussed in the conference, Lieutenant General BM Kaul suddenly realized that a major point had escaped

Mr HC Sarin, Joint Secretary - Ministry of Defence and the President, Indian Mountaineering Foundation and Bull Kumar's attention. The Barahoti Operation was colliding with his Nilkantha Expedition. After they missed Everest by 700 feet in 1960, the IMF had booked the Everest for 1962 and had planned three expeditions; Annapurna, Panchauli and Nilkantha. All these were virgin peaks. He was allotted the Nilkantha expedition and for which a lot of hard work had already been done.

However, the irony of the situation did not escape him as the Barahoti mission was overlapping with Nilkantha expedition's timings. And if he did not go on Nilkantha expedition, he would not be selected for 1962 Everest expedition. But he had no doubt in his mind that he was a soldier first and mountaineer later. Soldiering was his profession and mountaineering only a hobby. Therefore, the mission must take precedence over mountaineering. Bull who had been accused of poodle faking at the expense of his profession was now putting aside his first love to undertake that mission. He was not unduly worried about the Nilkantha expedition. It was sponsored by the IMF (Indian Mountaineering Foundation) and its Chairman Mr HC Sarin would be able to find a suitable replacement for him.

However, the Nilkantha expedition had not been forgotten by Mr Sarin. He told him that he was to return immediately after setting up the posts in Barahoti and hence be in time to lead the expedition to Nilkantha. At that time Bull Kumar thought he was the luckiest officer in the Indian Army. But fate had its own cards to play and time was to tell its own story.

The meeting finished on a positive note, General Kaul's optimism had rubbed off on the others because he was a true go getter. If he set his heart and mind on something, one could be sure that he would achieve it. He was a man of intense likes and dislikes. Although Bull Kumar thought he had full confidence in him and he was branded as Kaul's man, he made no concessions when it came to his likes and dislikes.

His views on the operation were solicited by General Kaul when he and Bull Kumar were the only ones left in the room. He asked him,

"What do you think of all that?" and Bull Kumar replied, "Nothing is impossible Sir. Only a lot of administrative details must be looked into. Equipping the party, getting the required rations and organizing porters would need a lot of time and attention." His reply to that was that the sky was the limit, as far as men, material and equipment requirements were concerned.

To begin with, everyone thought the mission was impossible if undertaken at that time of the year. Colonel (later Lieutenant General) Eric Vaz, for instance, who was the representative of the Eastern Command under whom the area fell, had expressed grave doubts about undertaking the mission at that time. Needless to say, he was not very happy about the fact that the mission was going ahead irrespective of his report. He had his reasons for feeling that the mission was being sent for some political reasons and asked General Kaul, "What are the orders for Kumar in case the Chinese attack him and his party?" A valid point no doubt, but General Kaul had all the answers and he replied, "Why think of a situation that was not going to arise?"

All the points raised by Colonel Vaz were valid - there was no way to send any reinforcements if the Chinese attacked and the Indian Army could not even assure Bull Kumar and his party of a safe escape route out of the area; but he was too young to understand the implications of all that. Like a true soldier, he just wanted to go to Barahoti and do the task given to him. The Kumaon Regiment training taught him to fight and not to ask questions.

They went to see the Defence Minister, Shri Krishna Menon, who too did not share General Kaul's confidence about the outcome of the mission. He suggested that we should go and call on the Prime Minister. That visit was no more heartening even though Pandit Jawaharlal Nehru was far more optimistic and keen about the mission.

Pandit Jawaharlal Nehru gave Bull Kumar one patronizing look and asked, "*Kya, aap yeh kar sakte hain?*" ("Can you do it?"). Bull Kumar sought his blessings and said that it could be done. All along, he had a feeling that Panditji wanted to give him a chance to back out. He probably thought Bull Kumar was too young to understand the difficulties of the mission and gave him the feeling that he was being nice to him. The Prime Minister asked him in all seriousness how early he could accomplish the mission. With confidence brimming, Bull Kumar told him that they could accomplish it by whatever date he told them. He replied, "He would let him know the date by which he wanted it to be accomplished".

Once the ball started rolling, the preparations moved at break neck speed. The Ordnance Depots all over the country were at Bull Kumar's command and they were told to cater to all needs. All the money necessary seemed to be at his disposal. He could make whatever local purchases he wanted with the senior officer with him detailed to look after the accounts. An Indian Air Force (IAF) Dakota was at his beck and call and he could pick up any officer, JCO or other rank for his mission.

He asked for the Para Platoon under training at Snow Warfare School, later renamed High Altitude & Winter Warfare School (HAWS). They knew little bit of skiing and were aware of mountain hazards like avalanches, crevasses crew, frost bites, high altitude sickness, snow blindness and were trained in mountain warfare. Bull Kumar had some kind of rapport with their members too. For a task like the one he had on hand, he wanted a team of people whom he knew and understood and who in turn had full faith in him to obey all his commands.

Captain VK Nair, the officer-in-charge of the 9 Para Platoon was Bull Kumar's friend and colleague from the Winter Warfare School. Affectionately known as Tubby, he was stocky and an extremely cheerful soul. That officious officer had all the guts required in a companion on a difficult mission like theirs. His second in command was second Lieut AS Cheema who played a vital role in the success of the mission. Later Captain AS Cheema was the first non Everest Indian Sherpa awarded the Padma Shri, Arjuna Award and Sena Medal.

When that Platoon came to Delhi from the snow bound mountains of Gulmarg, it was received personally by Lieutenant General BM Kaul and many of the top red brass from Army Headquarters Their patriotic feelings were being brought to the fore by the posse of distinguished officers from the Army Headquarters. The reception party shook Bull Kumar slightly. The members of the Platoon shared his feelings. Bringing to the fore the patriotism in all these soldiers was a noble gesture, but it should not have been done at the cost of diminishing the awe and fear of the unknown.

However, General Kaul and his colleagues from the Army Headquarters need not have gone into such an elaborate exercise to curb their fears. According to the best traditions of the Army, they were not only going to Barahoti willingly, but were even extremely enthusiastic about the expedition.

All that, while the preparations were getting along smoothly, during those hectic days Bull Kumar hardly slept, but strangely enough, he did not feel the lack of sleep during those ten days as he was too busy in organizing equipment and rations. He would fly in his Dakota to different Ordnance Depots all over the country to pick up the things he needed for his impending mission. At times he even took the plane to meet a few friends just to show off.

He was so short on time that he decided to take the easy way out when it came to making the food list. He just took the food list that had been made for the Everest Expedition and changed the heading to 'Barahoti Task Force Food List'. To his surprise it was passed without any queries. After 1962 War with the Chinese, that food scale was accepted as High Altitude Ration Scale. In later years, that act on his part was to prove beneficial for many senior officers and scientists of Defence Research and Development Organisation (DRDO). One senior officer realized that kaju (cashew nuts) was not essential on high altitude operations and by getting it removed from the list he saved the Army crores of rupees. One of the officers got juices cut and saved more money for the Army and for his efforts; he was awarded with the Vishisht Seva Medal. In fact, there were a number of officers of DRDO and Quarter Master General (QMG) Branch who owe their awards to that hastily prepared food list. All the items cut during the passive years in between 1962 to 1984 were reintroduced after the Siachen War.

Finally the preparations came to an end and Bull Kumar and his team were ready to leave with the advance party to Bareilly enroute to Joshimath. While at Bareilly, he had two free days. Here he learnt that a dining out party was being planned at Ranikhet for General KS Thimayya, Colonel of his Regiment, who was retiring as Chief of the Army Staff. Since he had some free time he decided to meet him. General Kaul got wind of Bull Kumar's plans and he was intercepted at Garam Pani and asked to return on his orders. The reason General Kaul gave was that he should not be distracted by such parties when he was on such an important mission. It was no secret that he did not like General Thimayya and so did not want Bull Kumar to attend his dining out.

All along the route the Uttar Pradesh (UP) Government had made arrangements for their swift progress. At places they were showered with such touching gestures and affectionate words that it was very difficult to hold back their tears that so often overflowed.

At every police post steaming cups of tea awaited them. Clean water for a wash was always there - an extremely thoughtful gesture after those dry and dusty journeys. Where the roads were narrow, the traffic was halted in advance so that they did not have to stop. With the help of a spare driver they drove nonstop for the next 24 hours.

They reached Joshimath the next morning, but Bull Kumar learnt with a sinking feeling that porters - the life line of such a mission, were not readily available. He had asked for 500 porters to transport stores, rations and ammunition. But his experience from his earlier expeditions (Trishul in 1958) in the area, he knew that was not going to be easy. He knew porters could not be got, no matter what the price one was ready to pay. He had come prepared, carrying with him a secret letter signed by the Chief of the General Staff.

That letter said that in case of emergency, he could use regular soldiers posted in that area to carry the loads. He handed the letter to the Brigadier in Joshimath who got furious but seeing that the letter had been personally signed by Chief of the General Staff he just raised his hands up in desperation.

In the 1960s, the motorable road only went up to Joshimath. Lance Naik Hansa Datt, Bull Kumar's man Friday from the Kumaon

Regiment was going to accompany him in the advance party. Dutt and Bull Kumar decided to trek to Reni, 20 km from Joshimath on the Dhauli Ganga River so that he could contact a few of the reliable porters who had accompanied him in his earlier expeditions. Even otherwise, Reni was known amongst trekkers going to Central Himalayan ranges of Nanda Devi and Trishul for getting reliable porters. In the village, he shortlisted two porters and their party of four moved ahead.

They trekked for 80 km spending a night each at Lata, Malari, Timmersen and Kala Jabar villages where they established the first Base Camp at the base of the difficult Chorhoti pass, situated at the height of 18,000 feet. Bull Kumar had decided to come in the advance party so that he could open that snow covered pass safely before any mountaineering expedition and he did not want to leave that task to anyone else. It took him and his helper four days to select a route and cut steps on steep slopes and fix ropes on the difficult portions for untrained porters and jawans to hold on to. That was necessary for the porters and jawans since they did not have any experience of mountain/snow climbing.

On one of the four days they were out there had been a huge avalanche which left behind massive destruction. Fortunately, his party was way up on the climb when that happened and had escaped its fury, but the ropes that they had fixed were not spared. They returned late in the evening to find no traces of the ropes. It was obvious that the avalanche had been a big one and they re affixed the ropes and returned to Base Camp after four days. The main party arrived while they were out and finding no trace of them they were in a panic.

Army Headquarters was desperately trying to locate Bull Kumar and when they got no news they panicked fearing what if the Chinese had inured Bull Kumar and his men.

Bull Kumar had left the wireless sets behind so that they could carry extra equipment for fixing the route. He felt rather foolish when they returned because the Area Commander had been severely castigated by the Army Headquarters for letting him go without the set.

There were various signals waiting for him. Some had advised, others had given good wishes but the one from the Medical Director was an order. They must acclimatize at 9,000 feet for 3 nights, 12,000 feet for 4 nights, and 16,000 feet for 2 nights before crossing the pass. On the morning of the D Day, Bull Kumar fell sick. He could not get up from his bed and suddenly felt a shiver run down his spine even as he realized that he was sweating profusely. The hectic activity of the last few days had finally taken its toll and his body had collapsed and what was worse at the most critical time.

High altitude sickness can start affecting the functioning of human body above 11,000 ft, and even one night's stay above that height, could make porters suffer nausea and headaches forcing them to return without their wages. Bull Kumar just could not take that risk and wanted to go straight for Chorhoti pass. After that going forward and returning would be the same. The porters at Chorhoti pass would rather go forward and earn their wages rather than return without them.

He almost burst into child like tears, as their doctor asked him to go in for complete rest for a few days. That was totally out of question as he was the only mountaineer in the group. If he did not accompany the party, the lack of confidence could be the cause of any number of accidents. If the porters slipped even once, the panic that it would cause would lead to mission failure.

He could not stay away. He told the doctor that he must go with the troops even if he collapsed on the way. Then the doctor told him what he could do to make things easier. Bull Kumar's body temperature was 104 degrees Fahrenheit and he doctor felt that he was planning, a suicidal mission. But seeing his determination he gave him some antibiotics.

The die was cast. Mountaineers say that no mountain - even the Everest – is worth giving up ones life for. But that was no mountain. That was a military mission, India had placed confidence in Bull Kumar and he could not – would not – let the country down.

Meteorologically speaking, the 23rd of April was the most favourable day. The westerly disturbances had just passed. The freshly fallen snow had consolidated enough to allow quick movement. Bull

Kumar issued some last minute instructions, took two sleeping pills and went to sleep. He was woken up by Hansa Datt at 11 pm. His sleeping bag was drenched with his sweat. Though he felt very weak the fever had gone - he just could not believe it.

He peeped out of his tent and what he saw gave him a fresh lease on life. Hundreds of torches were moving all over the place like busy little bees. But he could not even hear a whisper. Obviously, Captain VK Nair had everything under control.

It was a moonless night when they started out for Chorhoti Pass. He took the lead with instructions for the rest of the party to follow him after 15 minutes. They moved at a steady pace for about half an hour. Then he looked back at the progress being made by the rest. What he saw was breathtaking and enchanting enough to divert his attention for a split second. There was half a kilometer long line of torches weaving its way along the path he had carved out in the snow. The snow glittered and gleamed and reflected back most of the light thrown on it.

He had divided the porters in groups of 10. One soldier was kept in the lead and another one at the rear. However, it did not take long for reality to come to the fore. The lives of all the people were in danger and he could not imagine what the next few days had in store for them.

While opening the route, he had kicked steps all the way to the pass. Hence the progress was very quick. Only in a few places did they have to chop the steps again. They came to the first fixed rope that was not needed for a mountaineer; but for the troops who had not seen an ice axe before that was a life saver. For a slip there at that time of the night when the snow was hard and smooth, would take a person down for at least half a mile. There were 15 such fixed ropes to be crossed. Some of these had, of course, been fixed at very difficult portions.

Moving up in the dark his thoughts kept moving to Barahoti; what if the Chinese forces were waiting for them there? What would happen when their limited rations failed? Should they then surrender or fight and perish? They had enough rations to last only three days. He had been promised that food would be air dropped for them. But what was a promise if the circumstances prevented it being fulfilled? If the weather turned bad, and remained so, air dropping was going to be out of question. How long would it be before they would starve to death?

He was reminded of the question Colonel Eric Vas had asked General BM Kaul, "What were they suppose to do if they were attacked by the Chinese?" They were carrying 50 rounds of ammunition each. Even his young mind knew that it was not enough in case of an attack from the Chinese, but then some ammunition was better than none, he thought! If nothing else, it would act as a deterrent while they tried to find their way out.

After toiling for five and a half hours, they saw the first light of dawn that made the snow look dark grey that gradually changed to grey and as the horizon lit up to gold. Seeing such refreshing beauty of nature they forgot their exhaustion.

At seven in the morning, Bull Kumar reached the site of the emergency tent pitched to give shelter to those climbers who were exhausted by the first climb. When he had set it up, he had never imagined that he would be the first one to need its comfort. He went in and jumped into a spare sleeping bag. He knew the rest of the group would take another two hours to reach that point so he had some time to rest.

He got into the tent and prayed as he had never done before, both for the success of the mission and for the lives of the people with him. And then he felt guilty; why should he turn to God only when he had nowhere else to go? But by then sleep had taken hold of him and he slept soundly. When he got up, he wrote in his diary - "The moment I hit the air mattress, I went to sleep and dreamt of floating over the clouds in a kind of a craft. They soared up and up. Suddenly something happened. The craft overturned and started crashing downwards at a tremendous speed. Seven men got blown away by the wind and became invisible part of the universe. The remaining were shouting and cursing me. The shouting and cursing reached such a crescendo that it become unbearable and I woke up sweating and upset."

The first ray had already reached the tent. The porters were tired and downcast. They had enough of it and they did not want to go up any further. Even during the best of times, Chorhoti was a deadly pass. For the porters, who were very religious and superstitious, it was a pass ringed with mystery and fear. The locals believed that Chorhoti was the 'abode of demons and beasts'. There the ghosts of the departed were believed to hold black communion with the blowing winds. The evil ones were believed to keep lurking in the pass.

Fear stemmed from various legends attached to Chorhoti. The most popular of them had to do with a shepherd who perished in the pass with his sheep. It was believed that his spirit, and those of others who were killed there, haunted the area. Added to that fear, Bull's team also had to face severe cold that made walking difficult. Soon he had a mutiny on his hands; the porters refused to cross the pass. It was the temptation of double the wages that finally did the trick and they decided to move on. From there onwards, he had decided to move in the centre of the column.

They reached Chorhoti Pass at 11 am; the pass was bright and clear with the mountains forming a glistening background. They could see Kamet, Ganesh Parbat and other surrounding peaks. Scanning the heights, they could also see the immense blue emptiness of the sky. Below them lay the steep and mind numbing descent.

Lieutenant A S Cheema (who had climbed Everest in 1965) was in charge of bringing up the rear of the party. On reaching the pass, he found that some porters threw their loads and started walking backwards. So adamant were they in their decision to go back that when they were asked to stay put, they even volunteered to return the woolen garments that had been given to them, so that they could go back. Trying to entice them once again with higher wages too did not work.

Myth and rumours had convinced them that death was lurking somewhere in the pass and that they had a better chance of survival if they turned back immediately. These porters were carrying wireless sets that were absolutely vital to the success of the mission. If these sets were left behind, contact with the outside world would be completely lost. One of these wireless sets had already been dropped by a porter who cleverly dodged everyone and ran back. Cheema, the Jat Sikh, with only three months of service experience realized that. What Cheema did then might sound cruel in hindsight but it was what he thought the only, and the right thing to do. And till today Bull Kumar stands by his act. He brought up the deserting porters at bayonet point. He did not let any of them take a step backward. He threatened them with dire consequences if they did so.

The ascent to Chorhoti pass was quite gradual from Kala Jabar side. However it was extremely steep on the other side. There the ropes fixed by the team came in quite handy. It was a very long day for everyone. Later in the day as the sun went down they reached Rimkin where they decided to camp for the night. They were two kilometers away from Barahoti. From there onwards only the Revenue Party who was sent by the Indian Government to collect taxes from Tibetan traders were allowed to go forward for collecting taxes at Barahoti. Rimkin lay on a little plateau protruding outwards, formed by the lateral moraine of the glaciers that would have been there during the ice period, millions of years ago.

To save what little rations he could, Bull Kumar asked the porters to return towards Chorhoti pass so that they could cover the return journey the next day and gave them one day's ration. They relaxed a bit after the porters left. Their goal was well within reach. Barahoti was only a short distance away from them. So far, all had gone well.

He looked around; his eyes aligned on Hansa Datt of 3 Kumaon Rifles. In grateful appreciation, he shook his hand and half embraced him; for that man had done a lot! Like the porters he had carried his share of the load, he had held Bull Kumar's rope while he was negotiating the difficult pitches and when they returned to the camps while opening the pass. He would brew tea for Bull and cook meals. He slept late at night and got up long before anyone even stirred in the camp. When Bull was sick, he had stayed awake by his side. He had done all that without even a slight frown and not even a whisper of a grumble.

On 24 April, Bull Kumar had the best sleep in the last 20 days. Every one, of course, woke up on time the next morning but nobody disturbed him and he slept on till 1.00 pm. He got up in a state of confusion by the delayed hour. He stepped out of his tent holding on to his pants. The cramped tents made it impossible to tie one's pants inside. What he saw sent him staggering right back.

Lined up before him were the 80 porters he had sent away to spend the night at the base of Chorhoti pass. By now they should have been on the other side of the pass, well on their way home. Why were they there? What happened? What had caused them to return? His immediate reaction to the situation was – how were they going to feed these 80 men for one more day? Twenty four hours was the minimum time required to go back to the Base Camp and replenish their stocks; they had left 30 ration loads at the Base Camp and barely had enough food to last two days.

Captain VK Nair came to him for consultation and advice. It was obvious that they would have to feed the 80 tired and hungry porters. His serenity and tranquility seemed to abandon him. Shaken and agitated Bull Kumar started thinking ahead. Soon, very soon, they would all be starving. Their only salvation was an air drop of supplies. But would it be possible for planes to drop food near the border, in such a narrow valley? He decided to cut down on the rations of the entire team of eating wholesome breakfasts of six puris and sabzi and 10 chapatis each for breakfast, lunch and dinner.

The soldiers had no choice but to stay there and remain hungry till the mission was accomplished, but he felt it was unfair to keep the porters back. He asked Captain Nair to give them enough rations for a day and send them back with a warning that they were not to return.

For the rest of them the orders were clear; move, and move fast. He wanted to contact the Army Headquarters in Delhi for help but he wanted to do that only after announcing the success of the mission. With Barahoti so close, he first wanted to accomplish the mission and subsequently bring up issues like food and hunger.

It was co-incidental luck that Head of the Revenue Party later became Bull Kumar's *'Samdhi'* (Relative) as his daughter was married to his son. These men had come to establish and verify for themselves their civil rights to the disputed area. Bull Kumar asked them to be ready immediately as unless they could place themselves beyond the dispute on Barahoti, the success signals could not be sent. Once they had placed – absolutely and physically - Indian civil administration on Barahoti they would inform Delhi.

They trekked the last 2 km to Barahoti and immediately sent a signal to Delhi. Their success signal was the 'battle cry' of the Marathas –**'Chhatrapati Shivaji Maharaj ki Jai'** as these were the troops with the mission.

He learnt later that the staff officers at the Army Headquarters in South Block were in a quandary over the contents of the signal. It was some time before they could figure out that the message signalled success of their mission. Bull Kumar immediately received personal messages of congratulations from the Prime Minister, the Defence Minister and Lieutenant General BM Kaul.

The success of the mission might have been a source of rejoicing in Delhi, but their troubles were far from over. Now they had to undertake a struggle for survival – not necessarily against the Chinese forces but against the most basic need - food for survival. His success signal was followed by an SOS flash- "**starving, request air drop food.**" And then they waited for a miracle to happen.

The 25th of April, the third day of the expedition, dawned bright and clear. On their half empty stomachs its beauty largely went unappreciated. The moment they had issued the extra rations to the porters two days ago, their rations had become very meagre. Even the emergency rations that they had carried were by then exhausted.

The 25thApril was the last day for which they had food and their frugal lunch would be their last meal till they got more stocks. So badly were they placed that even if they wanted to retrace their steps to reach some village, they could not; Timmersen was the closest point where they might get food left by the porters, but it was four days away with the dangerous Chorhoti pass in between. Habitat villages like Ghosali and Malari were 8 to 9 days away and their depleted energies could not have withstood the pressure of the arduous, cold and the demanding journey.

Hunger, cold and insecurity had hitherto been abstract terms for most of them; but to understand their real meaning is an entirely different affair. It is only when one suffers personally that one can understand the experience and full import of having to go hungry. Furthermore, they were experiencing that lesson in the solitude of the snow covered mountains of the mighty Himalayas. But it is an equal truism that man's endurance is truly shocking! Dr. Ao proved that amply. Belonging to a tribe lived in the wilderness of Nagaland - Dr. Ao was the happiest amongst all when he left civilization behind. Their natural instincts for survival became sharpened as they moved higher and deeper in the snow covered mountains.

Dr. Ao approached Bull Kumar in the evening and in hushed whispers tried to explain a plan. His accented English was difficult to understand at the best of times and on an empty stomach with no signs of help at hand, Bull Kumar had a real problem trying to fathom what he was saying. But when he understood his plan he sprang out of his bed, put on his clothes and followed him as a man possessed.

Dr. Ao confidently moved towards the Chinese border with him following expectantly. They turned a corner and a startling sight met his eyes. Eight big rats lay dead on the snow. That was the treasure that Dr. Ao had unearthed on his own and wanted to share that with the rest of the team.

The moment Bull Kumar saw the rats he realized that Dr. Ao had spent the whole day improvising traps out of nothing to catch the rats. His noble thoughts, notwithstanding, the thought of eating the rats made his stomach churn. Dr. Ao however, assured Bull Kumar that the meat was healthy and in his part of the country was also a luxury and delicacy. The Brahmin in Bull Kumar revolted at the thought of having to touch these filthy creatures. But as beggars could not be choosers, he started helping Dr. Ao in skinning the rats. They then chopped them into pieces that were later cooked and served as deer meat. Dr.

Ao relished the dish. So did the others. Bull Kumar was the only one who did not enjoy the delicacy. Back in his tent Bull Kumar vomited what little meat he had eaten.

They woke up on the fourth morning without any early morning tea or breakfast. Whatever hope they had so far was dying and they were all beginning to give in to despair. He knew he could not let that happen so he made all of them come out of their tents and sit and talk. They argued and reasoned and he tried to convince them that Delhi would not forget them. Finally, he too started disbelieving what he was telling them. He also tried to play on the desire in each one of them to live. He knew that once the men lost interest in surviving, it would be difficult to keep a hold on them.

In the meantime he tried to figure out how long it would be before the air drop. The previous day, he figured must have been spent preparing and briefing the pilots who were to come to their rescue. Today, he felt, was the most important day as the Dakota's would attempt the drop.

Sitting despondent and hungry, they suddenly heard someone shouting, 'He could hear that was coming. He could hear that too'. He must have had very sharp hearing because nobody else had heard anything. A wave of excitement ran through the group. Finally they all heard a faint buzz. For a while they thought their imagination was playing pranks with their minds. But the buzz gradually became more distinct. Bull Kumar stole a glance at Tubby, he smiled back, calm and unperturbed. The gamble was producing results. It had been worthwhile. They had already marked out the dropping zone (DZ), and kept the smoke candles in place. A huge 'D' was spread out next to the dropping zone. Their recognition sign was 'Delta' for the DZ.

They looked towards Grithi gorge from where they expected the planes to come. The Dakota appeared like a huge bird soaring gracefully through the narrow gorge that brought with the hope of survival. Captain VK Nair had recently done an Air Ground Support Course and so he was in charge of handling the air drop. He got his wireless set and tuned on to the agreed frequency. Anxiously he spoke into the set, "This is Delta DZ. Delta DZ. Delta DZ calling, report my signal, over." Then he waited for the pilot to reply but the response was deathly silence. There was no contact even as the Dakota came nearer. It came very near and then, to their dismay, turned into the adjoining valley.

They looked at each other trying to figure out what that meant in utter dismay and disbelief. They had only three smoke candles with them. They had left the rest behind because they were quiet heavy. One of these had already been burnt to guide the pilot. In desperation, Bull Kumar asked Cheema to climb onto a nearly peak with a smoke candle and light it there. Perhaps the pilot would be able to see it if that was placed at a height. In a trice, Cheema charged up a 16,000 feet high ridge with amazing speed.

The plane appeared once again, emerging from the adjoining valley. Heartened they burnt another smoke candle. And then they removed their heavy jackets and began waving them in an attempt to catch the pilot's eye. They even organized a collective shouting system. If nothing else, Bull Kumar helped his team to give vent to their fears and frustrations.

The Dakota took a very sharp turn into Grithi Gorge. Its wings seemed to almost scrape the mountain side and then it went back into the depths of the adjoining valley. Although they admired the pilot's guts for flying in such a difficult mountain terrain, they also cursed him for his despicable and scanty navigational sense. Nair even swore to give some map reading lessons to the entire crew if they made it. He was busy shouting into wireless set desperate to make contact.

Three times they saw the plane disappear into the adjoining valley only 500 yards away from them. The last time the plane disappeared from the valley they saw the smoke blazing from the nearby ridge. Cheema had lit the last candle at just the right time but the pilot failed to see even that. Their hearts sank when they saw the plane turning back. Bull Kumar Kumar overheard one of the men say, *"Ab to mare gaye, ab kya hoga?"*

With the passing of every moment the plane kept drifting away and diminishing in size up in the sky and so did their hopes. That really looked like the end. There was still one meagre hope left though. They were till banking on their wireless set to establish contact. The operator was repeatedly crying, "Delta DZ, Delta DZ, Delta DZ, Delta DZ, report my signals, over." They all turned their attention towards the wireless set, hoping for a miracle.

As the Dakota was disappearing fast, panic started gripping Bull Kumar; in case of a Chinese attack they had ammunition for just a couple of hours. Moreover, if someone fell seriously ill they had no means to help him. While the officers were all bachelors, with not many responsibilities, the troops and members of the Revenue Party had other responsibilities. They had children to look after, dependent parents and wives to support. The future of all these people depended on the pilot.

Bull Kumar was willing to compromise on any issue except retreating from Barahoti. He still had faith in the Army Headquarters. That faith was restored when his thoughts were broken by a feeble voice coming from the set, "Where are you Delta DZ"?

Captain Nair took charge immediately, "That is Delta DZ Special Task Force. We were starving. You must come back. Over"

Now that the contact was established the pilot told them his problem that there was too much glare from the snow and he was unable to locate them.

The indefatigable Tubby was not to be suppressed by that. "Hello, come back and I shall guide you." The pilot said that he was running short on fuel and he would come the next day. Captain Nair would have none of that. He was determined to get the plane back; "Whosoever is flying the plane listen to me. We have eaten nothing for two days. You have to come back today".

Hearing that, the pilot turned and they started hearing the steady drone of the plane. In less than 5 minutes it was flying over them. The pilot was flying so low that they could even see the open doors of the Dakota. The pilot still insisted that he could not see the DZ markings; but the calm and confident Captain Nair guided the plane in circles over them and told the pilot to drop supplies on his orders.

Thus began one of the prettiest operations ever witnessed by them. The food fell out of the plane in sacks and splashed in the wet snow. No other sound could have been more thrillingly welcomed than the noise of the falling supply bags. The heavens had literally opened for them. Bull Kumar looked around and saw some Jawans kneeling on the snow and praying and tears of gratitude were flowing down their cheeks. The plane unloaded 57 bags after that the pilot refused to drop any more stuff. He was running so short on fuel that if he hovered around any longer he would never make it to the base. He had more stuff for us that he promised to deliver the next day.

The recovery of the loads was carried out with great enthusiasm. Cheema saw a goat running away from that area where the load had been air dropped. He ran after the goat to retrieve the animal. But that turned out to be a wild goat, that made Cheema climb faster and faster before finally disappearing.

They immediately started preparing the first meal in the last two days. While there were enough supplies they realized that the bags of cooking oil and salt were missing. But who was going to complain of luxuries like salt when they finally had something to eat! A sense of happiness enveloped the camp as they sang and danced around the camp fire.

They all woke up early the next morning to get ready for more supply drops. At around 10 am they heard the familiar sound of the Dakota again. It became loud and clear even though it was yet very far away. Since they had run out of smoke candles they instead got some boxes to burn to attract the pilot attention.

Suddenly the sound of the Dakota stopped and there was deathly silence in the mountains. Bull Kumar froze on his feet. Perhaps they were mistaken and it was not the plane they had heard after all. They waited for something to happen, but nothing did. They did not hear the plane again. It was unbelievable and frightening.

He turned to Dr. Ao who was sitting next to him and asked him if he had really heard the sound of the plane. He was hoping that he would say no, but Bull Kumar's hopes were in vain. They all had heard the plane noise just as them followed by the deathly silence afterwards. Bull Kumar had a gut feeling about what had happened but he did not dare to speak out what he felt.

Their next contact with the outside world was at 12 noon. The message they got read, "Dakota missing". Bull Kumar's fears were confirmed. The air craft carrying supplies for them had crashed. Even though he was sure that the plane had not crashed on their side of the pass, he sent their patrols to comb the area. 2nd Lieutenant Cheema's patrol returned very late. As there was no demarcation of borders Kumar hoped he had not been captured by the Chinese. The tireless Cheema had walked for about 12-13 hrs without any success.

In the evening they confirmed to the Army Headquarters that the plane had not crossed Chorhoti Pass. What distressed him the most was the fact that the crew had risked their lives to keep them alive and they were helpless to do anything for them in return.

On his fifth day at Barahoti, he decided to return. He had done all that was in his power. They had accomplished their mission. The men also had a stock of supplies and from now on their duties and destinies and his lay in different directions. He would leave Barahoti with a small team and the rest of them were to stay at Barahoti under Captain VK Nair. Bull Kumar had another task ahead of him waiting - the Neelkantha Expedition! When he announced his intentions, he could sense some bitterness amongst the men. They were duty bound to stay and protect the borders, whereas he had the freedom to leave.

It was depressing to take leave from the team. They had become very close to each other as they had faced one challenge after another. The camaraderie and bonhomie developed in that mission was emotionally very intense. They had also come to understand and anticipate the slightest change in each other's moods. It was difficult to break away from all of them.

Next morning at 8:00 am, Bull Kumar left with Hansa Dutt and seven soldiers from the Dogra Regiment. He left Barahoti and Rimkin with an empty feeling. Half heartedly he reached Chorhoti Pass but even that challenging ordeal could not get his spirits to rise.

The zest and interest of an unaccomplished task were no longer there to goad him on to a faster pace. The journey homewards, he knew was comparatively easier than the journey outwards had been. He felt tired and listless. For the first time since he had left Delhi, he became concerned about his impending expedition to Nilkantha.

Before tackling the very steep climb of 1,500 feet to the top of Chorhoti pass, he decided to take a break over a cup of tea. He remembered a huge overhang of rock and decided to rest there. The moment he reached the rock, he heard soft sounds of the heavy breathing and moaning. Since no human being was supposed to be there he assumed it was an animal resting under the shelter. He approached the place and could not believe what he saw. Under a black blanket lay two skeletons - immobile and inert. As he came closer, he realized that the skeletons were still alive.

They were obviously the two porters that he had sent back from Rimkin four days ago. It was a miracle that they were still alive. Their legs were black up to knees due to severe frostbite. He tried removing their boots, but the boots, socks and flesh were cemented to each other. It was a sickening sight. One of them had enough strength to speak. He mumbled, "*Sahib, mujhe bacha lo, meri bibi ka aur koi nahi hai*" (Sir save me, there is no one else to care my wife).

The two porters had become victims of hypothermia - the general effect of cold accompanied by symptoms of tiredness and sleep. Once a person gives in to that weakness it gets very difficult to overcome. It gives one a feeling of having had a couple of drinks and to stay in that warm condition. The victim keeps postponing the thought of getting up till it gets too late and he freezes to death.

Bull Kumar thought and felt death due to hypothermia must be one of the most peaceful deaths one can have. Though he was alive, he had the experience of that comfortable feeling till he was shaken out of it by his colleagues. These two must have decided to rest for the night before crossing the pass; by doing so they must have come under that unavoidable and strength sapping hypothermic coma.

The question foremost in his mind was how to get the two of them on top of the steep pass. Once they were at the top of the pass, he knew there would be no problem as he had left a ski sledge there for such an emergency.

They cleaned the two men who had been lying in their own filth for four days, gave them a few drops of water and two men team carried them for the first 20 yards. At that height it was difficult to even carry an extra ounce and carrying two human bodies was proving to be next to impossible. The track was so narrow and steep that they could not carry them on stretchers.

Soon the first team tired out and the second one took over. After covering a few yards one of these soldiers sank to his knees and started vomiting. He was immediately relieved of his burden. But the sight of blood in his vomit completely demoralized the rest. The Havildar camp up to Bull Kumar and said in an accusing tone, "*Kya hum in coolion ke liye apne jawanon ki jaan de denge?*" (Should our soldiers die for these coolies?)

Obviously all of them were agitated at Bull Kumar's decision to carry the two porters with them. The underlying tone of the Havildar's voice was accusatory. He was blaming him for the problem they were in. "You have landed us in that predicament and risked so many lives", his tone hinted. How could Bull Kumar tell him and the others that that was not so? Their feelings were just getting carried away by the despair they were facing at the moment. How could he tell them that he was only following orders and had no personal stake in the Barahoti!

The soldiers were talking and whispering amongst themselves. One of them, more tired than the others, started crying that he would never reach home. That was fast turning out to be Bull Kumar's most critical test of man management skills.

On the one hand there were the two porters who had to be given a chance to live and on the other was the vital question - should they risk the lives of the soldiers to save them? Without saying a word to the Havildar he got up and walked towards the men. He addressed them in a clear voice,

> "*Sathion, shayad aap ko malum nahin hai aap kis oonchi battalion se taluk rakhte hain. Hindustan mein sirf ek hi aisi battalion hai jis main ki ek VC (Victoria Cross) abhi tak serve kar rahe hain. Shayad aap me se jo naye recruit hain unko yeh bhi malum nahin ki Indian Army, maidane jung*

main kabhi bhi apne zakhmi ya murda aadmi ko bhi nahin chor karaati, jo aaj aap karenge yeh sari umar tak aap aur aapki battalion ke liye gorav ya sharam ki baat rehegi".

(Friends, perhaps you are not aware as to which battalion you belong to! In India there is only one battalion that has a serving VC(Victoria Cross). Perhaps young recruits within you are not aware that the Indian Army never abandons its wounded and dead comrades in any battle. Your actions today today would either bring pride or shame remembered by the posterity.)

It was the call to the highest and the most noble in their nature to rise to the occasion and behave like the brave men that they were. Having said that he beckoned Hansa Datt and together they walked towards the casualties. He picked up one of them and Hansa Datt the other, and they started the uphill climb. Thereafter, every step seemed to be the last one of his life. Bull Kumar's lower limbs shivered under the heavy weight. The fibers in the muscles seemed to stretch to the maximum and came apart at the calves. His perpetually wide open mouth sucked in gulps of air, but even that was not enough to meet the need for more oxygen. Sweat broke out on his forehead and he felt as if he was about to collapse.

A little breeze from the east brought relief and gave him a breather. His eyes began to see clearly and he began to feel better and stronger. He put the porter on the snow and stood up to have a look around. He realized that he had merely covered 20 steps even though it felt as if he had been climbing up for the entire day. He was heartened to see that his words had the desired effect on the soldiers some of whom were already lending a helping hand to Hansa Datt.

Soon two soldiers came up to him and together they carried the two porters to the top of the pass. Just as they were nearing the pass, the Havildar came and tapped him on his shoulder to halt and told him, "*Yeh to mar gaya hai*" (he is dead). Bull Kumar only then realized that he was carrying a dead body. For the first time in his life he had been a pall bearer under most adverse circumstances.

They tied the living and the dead porter together to the emergency sledge and started their journey downwards for Kala Jabar. The climb down was a lot easier but now there was the danger of avalanches. However, being in extremely tired physical and mental state, the implications of being hit by an avalanche were not that scary. At Kala Jabar they first went to the post and completed the panchnama. That was the local custom requiring five witnesses to declare a man dead before cremating him.

From there the track was much better. They put the living porter on a stretcher and headed for the Company post at Malari. Their Dogra companions were engulfed in affectionate embraces by their friends. It was with great amusement that Bull Kumar overheard some of the horror stories that they related to impress their friends in the Company.

He was escorted by the Company Commander, Major Ajmer Singh to his tent. Here he had one of the most memorable and delightful treats of his life – a salted parantha, a meal that had never tasted so good before. Having foregone salt for two days, he realized for the first time how precious salt was for living.

Major Ajmer Singh unfolded to Bull Kumar the story of the Dakota crash. The plane had come in the morning flown comfortably over Malari before getting caught in an up drift. Miraculously, there had been four survivors. Emotively Major Ajmer Singh showed him a telegram from the navigator's wife, "There is no war on. Can not you even give us the ashes of our beloved one?" Bull Kumar was greatly moved by that appeal. He looked enquiringly at the Major for the remaining story. He explained that ten patrols had been sent in different directions to locate the crash site. But they had all been unsuccessful. Various aircraft too made a number of sorties in the area for three days but could not locate the crash site.

Bull Kumar was so moved by the telegram that he decided to track down the crash site. After all, these people had lost their lives in providing food to him and his team. The least that he could do for them was to give them appropriate after death obsequies. He made that his mission.

Bull Kumar asked Major Ajmer Singh to supply him with some man power. He told him that he had to bring the ashes of the dead.

Major was stunned and when the shock was over he tried to dissuade Bull Kumar from undertaking his mission. After all, what could he do when ten capable patrols, helicopters and reconnaissance planes had failed to locate the wreckage?

Bull Kumar pointed to the telegram and told the Major that he felt morally bound to bring the ashes of the dead souls. And he had to start in a hurry because his Nilkantha expedition was to take off just as soon as he got back. But he wanted to get the ashes of the dead personnel before that as that would satisfy his conscience. Having led the mission to Barahoti, deep down Bull Kumar blamed himself for the deaths of these people.

Seeing his determination, Major Ajmer Singh agreed to accede to his demand. But there was still one difficulty before the men could be detailed to help him. He had already told his CO that the patrols sent out by him had failed to locate the wreckage. That difficulty had been further confirmed by the four survivors. They had even said that it could be dangerous to carry on with a search. With all the advice given to him, the CO had declared the chapter closed. Reopening the chapter would require his consent and permission. And that could not be achieved before 9'o clock the next morning. That would be far too late for him because by then the spring sun would be out softening the snow. That would be an impediment to swift movement so he told the Major that he wanted to leave Malari by midnight. He looked at him and told him that it was already 7:00 pm.

Though Bull Kumar had many volunteers who agreed to accompany him, the Company Commander could not allow them to go without orders from Joshimath. Luckily for him, he came across a labourer with the Company. Bull Kumar took him aside and even before he could ask him to accompany him, he agreed. Then he looked guiltily and questioningly at his faithful Hansa Datt. There was no hesitation in his eyes.

Having his team in place he got busy sorting out the stores he wanted to take with him. He decided to take the minimum possible; a stove for cooking, a small jerrican of kerosene to cremate the dead and 40 chapatis to live on along with one small tent for the three of them, a climbing rope, some pitons, a hammer, three feather coats but only two mattresses. He decided not to take any sleeping bags, for they would be too heavy and cumbersome to carry. He thought it was best to spend the night in their feather coats.

He did not want to leave behind his camera. He also took a khukri that he considered a necessity for cutting bushes. To that lot, he added some torches with spare cells, two flasks of tea, three water bottles and some dhoop (agarbatti) borrowed from one of the Jawans. He asked the sentry to wake them at midnight. As he tried to go to sleep his mind kept returning to his fatal dream in that seven people had drifted away from the vessel and they had lost six lives from his team. He prayed to God to spare them any further deaths. He asked for Divine mercy.

He was fully awake when it was time to leave. Minutes later their three torches had moved into the dark night. How small and insignificant that caravan looked as compared to the one that had left Kala Jabar for Barahoti only a few days earlier. The first three miles were easy as they were moving on the same route that they had covered earlier in the day. While approaching Malari earlier, he had never for a moment thought he would be retracing his steps so soon. They crossed the Girthi Nala and then followed a beautiful mule track for another three miles along the right bank of the Nala. The track ended abruptly into a foot path.

He had studied the map thoroughly and had selected two big Nalas that were wide enough to have accommodated a big plane like the Dakota. It was still dark when they reached the first Nala coming down from the ridge south of Kala Jabar. He called a halt. Each one of them was carrying about 30 lbs of weight and was tired by then.

There was another reason for that break. There was no track from there onwards. Hence it would be foolish to carry on in the dark. There was still about half an hour before day break. They used that time to recuperate. The moment they sat down, his companions dozed off. Dawn provided them the light they needed to move on. They felt greatly refreshed after a short sleep and scrambled up the steep climb venturing into a thick patch of rhododendron bushes. Hansa Datt showed great skill in cutting out a patch through the thicket.

Badly inflicted with scratches, cuts and superficial wounds they managed to cover that stretch in four hours. The snow was very wet and soft under the trees and it was with great relief that they emerged into the clear spot feeling very satisfied with firm snow under their feet. They ate while climbing and on the spur of the moment Bull Kumar decided to climb to the left hoping that would provide them with a clear view of the entire valley. Unfortunately, the view ahead was badly limited. The only thing they could see clearly was the Dhauli Ganga flowing far below them. He reckoned that by now they had climbed up about 1,500 ft but still there were no signs of the crash.

The climb beyond that point was tougher. Their legs gave way and their arms ached due to the pressure of the rucksacks. The straps were acting like tourniquets preventing proper flow of blood. At noon they took a break, left their stuff and carried ahead with only oil and climbing equipment. Having got rid of the extra load, they made brisk progress. On the right ridge he glimpsed a shining object. Hoping that he had spotted the crash he climbed the ridge to inspect it only to discover that it was a piece of dry wood. How could he have been fooled so easily?

By 2:00 pm, he had started wondering whether he should give up the search. In that Nala, there were only a couple of thousand feet of climb left to the top of the ridge. A huge crash, if it had occurred there, could not have remained hidden from them for so long. After resting for a quarter of an hour, they left the tin of kerosene oil and decided to climb the bump above them just to make sure, and confirm for themselves, that no area had been left unsearched. Knowing that he had already done a thousand extra feet, Hansa Datt and Thapar Singh very thoughtfully took the initiative and made the track for him. He mechanically dragged himself in their footsteps, leaning on his ice axe trying to relieve his exertion.

Suddenly he heard a loud shout from Hansa Datt, "*Sahib Mil Gaya*" (Sir, found it). The crash site had been found. He saw the moment he got up to the top of the bump that lay sprawled just below them. It had remained hidden from their view by that protrusion in the ridge. Just as that projection had been shielding the wreckage, whose unexpected presence had perhaps also been the cause of the accident.

The wreckage was spread all over the area, just fifty feet away from the ridge. The front of the plane was totally burnt, while the rear was intact. The wings had been shattered to bits and the debris lay scattered all over the place. Bull Kumar stared at the sight for quite some time, too scared to approach it. He felt as if the dead would arise and question his presence. With a sickening feeling in his stomach he tiptoed to the door of the plane, but could not muster courage to enter and jumped out almost immediately.

He braced himself for the second time and asked Hansa Datt to get the kerosene oil. His moment of fear and turmoil had passed and he no longer experienced any remorse for the dead. He had a mission to accomplish and that was the reason he was there.

He drew a chart of the wreckage and tried to identify the bodies. Near the door was a half burnt body with a completely charred face. It reminded him of the frightening characters in a Hitchcock film. His badge said he was a Junior Commissioned Officer (JCO) of Army Service Corps (ASC). He was wearing his khaki Army jersey. Seeing him he was reminded of his elder brother who was then in the ASC. Crushed among the bags of dal was a Non-Commissioned Officer (NCO) fully preserved. He looked in his pocket and found a letter giving his name and identification.

In the navigator's cabin he found two completely burnt and charred bodies. One body had a microphone around his neck. That he concluded was the signaller. He emerged from the remains of the navigator's compartment and moved over the debris that had been the co-pilot's seat. A charred body had his legs struck sideways in his seat.

Like a machine Bull Kumar collected souvenirs, took photographs and completed his diagram of the wreckage. By then, Hansa Datt had returned with the kerosene oil. Collectively, the three of them dragged out the five bodies, collected burning material from the wreckage and built a joint pyre for them. They offered prayers, poured kerosene oil over the pyre and then lit it. Bull Kumar stood back and watched the ashes being formed. These were the ashes he had sworn to take back with him.

He thought of the only pyre he had ever seen before. It had been his grandmother's. She had died when he was eight. He could still recall the way she had been given her last bath and had been carried to the cremation ground. He thought he had loved her dearly and all the noise and the chanting of prayers had given the ceremony an impersonal feeling. This occasion however, made him feel sadder at the loss of lives. This incident left a far greater impact on him than the death of his grandmother. When the bodies were fairly burnt he collected the ashes, placed them in three separate bags, labelled them and tied them firmly. He once again looked over the wreckage of the plane. That was the plane that had set out to bring them food.

Satisfied and at peace with himself, they started back for Malari. The sun was way down when they started back and he decided to leave behind the equipment they had brought with them to enable them a quick journey back with the ashes of dead ones. All the same, he picked up the climbing rope as a precautionary measure. The journey back was fast and they reached Malari at 11 pm. Everyone in the company was surprised to see them back so soon. What was even more unbelievable that they had taken just one day to locate the wreckage and bring the ashes.

He was about to retire for the night when a signal was shown to him. What he saw shattered his sense of peace. One of the soldiers at Rimkin had died of high altitude sickness. How costly was his dream coming to a completion! The seventh man of his expedition had died. It was a heavy toll to pay for one mission.

The next morning saw him up again, agitated within, but controlled outwardly. He walked 42 km to Joshimath in 7 hours that helped him to give vent to all the tension, frustration and angst that had built up within him.

In contrast, the atmosphere in Delhi was one of gaiety and celebration. Everyone in the Army Headquarters was elated with the successes of Bull's mission. He was accorded a hero's welcome and taken to the Defence Minister and then to the Prime Minister for a pat on his back. They were indeed liberal with their words of praise and congratulations. It seemed strange to him - he had lost seven precious lives, and yet he was being patted on the back. During his first meeting with General Kaul, he expressed regrets about the deaths. His reply left Bull Kumar shell shocked as he said, "The work that you have accomplished was worth 700 lives".

After the mission, he uncomfortably became important overnight. Even though he should have taken the Nilkantha expedition immediately, he was not allowed to proceed and was dismayed. Of all the things, he was being held back to write a report and could not go for another adventurous and dangerous mission. Permission would not be given to him for the Nilkantha expedition.

He discovered that the pilot of the crashed plane – Wing Commander De Soares was in a hospital in Delhi. He decided to meet him and he told him how the Dakota had crashed. During the flight, he had followed the Dhauli valley that brought him over Malari. From there he had turned right into the Grithi Gorge. But, before he could complete that turn, the up draft had taken him up and forward. He kept climbing as the aircraft approached the mountain and then suddenly the air craft would climb no more.

The mountain loomed close, large and formidable. He had no choice but to force land on snow. He was lucky that 20 feet of packed snow had provided him with a good shock absorber, and at least some people had managed to escape. At the wreckage site he had noticed a huge rock adjoining the spot of the crash. Bull Kumar was curious to know how De Soares had managed to avoid it. Good humouredly came his response, "The pilot was an experienced man". But the pilot on the other hand was surprised, at to how Bull Kumar and his team had climbed up to the crash! On the contrary, they had just slid down holding bushes and reached the lower reaches with almost bottomless trousers.

NILKANTHA - 1961

Bull Kumar was allowed to leave for his Nilkantha expedition only after a month, at the end of the May. The expedition was a success and they were the first Indian team that conquered the peak after six failed attempts. Nilkantha (21, 640 ft) is one of the most beautiful and difficult peaks in the Garhwal Himalayas. Prior to the ascent of that peak in 1961, it had defied six foreign attempts including those made by Frank Smythe, Willie Unsold and Sir Edmund Hillary. Though, that peak has the easiest approach march in the Central Himalayas from Badrinath it had the reputation of being an unclimbable dangerous giant.

The first task was to select a suitable team. In the mountains, close teamwork is of supreme importance and Bull Kumar would, therefore, prefer moderately good, but a well knit team of climbers. He was able to get together a group of strong climbers of whom he was confident and who would work as one motivated entity. The expedition team consisted of six members. One was Flight Lieutenant AJS Grewal who was a former member of the Indian Everest expedition 1960, and was popularly known as 'Khalifa' in mountaineering circles because of his ready wit. He was the only married member of their team.

There was Flight Lieutenant A. K. Chowdhary also a member of the Everest expeditions as well as of the 1957 Nanda Devi expedition. He climbed Choukhamba (23,420 ft) in 1959 and was an asset to them as he had also been a member of the IAF Nilkantha expedition. He was known as "Ang Chowdhury" for his youthful charm. Shri OP Sharma, the third member was an extremely keen mountaineer; he had done a course at the Himalayan Mountaineering Institute. Tenzing had said of him that he had the makings of a good climber. Being the youngest member of the team Nilkantha was to be his first major expedition.

Lieutenant RC Roy of the Army Medical Corps was the expedition doctor. This was to be his first experience in the mountains and he was keen and enthusiastic. He took his job seriously and compelled everyone to swallow the prescribed pills at meals. The last member of the team was Captain Badhwar, who was a professional soldier.

Captain Mulk Raj, whose name was not on the team, was very keen to go on the expedition. Bull Kumar knew him from the IMA days as he was a year senior to him. So Bull Kumar informed him that the team was complete but if a vacancy were to arise, he would include him. The two would go out trekking together as far back as 1950 when they were cadets at the Indian Military Academy, Dehradun. Captain Mulk Raj had maintained his interest in mountaineering and had been a member of the 1959 pre-Everest expedition as well as of the late Major "Nandu" Jayal's expedition to Nanda Devi in 1957.

When Captain Badhwar was told by his CO that instead of wasting his time on the expedition, he should concentrate more on his profession, he decided to drop out of the expedition. Bull Kumar then told Captain Mulk Raj that he could join the team in place of Captain Badhwar. Being the only vegetarian in the team, he was so keen to go on the expedition that he promised to become a non-vegetarian during the expedition. So, while going up to Base Camp, he went to Badrinath Temple and asked for forgiveness to God for eating meat during the expedition. Actually it turned out later that he relished non-vegetarian food.

So Bull Kumar's final team consisted of Flight Lieutenant AK Chowdhary, Shri OP Sharma, Flight Lieutenant AJS Grewal, Captain Mulk Raj and the Doctor, Lieutenant RC Roy. He had requested Tenzing to select five Sherpas for their expedition and selected a fine set of men. Three of them had carried up to the South Col on Everest in 1960; and one beyond to Camp VII. They were Phurba Labsong, Lhakpa Lama, Nawang Tshering, Tashi and Pasang Lhakpa.

Of these, Phurba, who was extremely good at rock-climbing and ice-work was the wittiest of all. He was the life of the party. Devoted to his profession, he looked after the rest of them like a fond parent. He was also a daredevil, full of courage and pluck. He would gladly stake his life for the sake of his companions. Phurba who met his death soon after the expedition was noble and brave to the last. He died on a German photographic expedition to Pumori only a few months after he returned from Nilkantha.

With the Everest Expedition still fresh in his mind, planning was not difficult. The HMI, Darjeeling had loaned them most of the equipment they required from the Jayal Memorial Stores. The Ordnance Factories manufactured the rest while some was loaned by the IAF trekking Society. Bull had assigned Grewal and Roy to manage the food arrangements and medical stores respectively. Sharma handled the donations and Chowdhury took care of the porters. Bull Kumar took charge of the equipment.

The party reached Badrinath on the 30th of May. Previous expeditions had tried the mountain from almost all conceivable approaches. Therefore, he had immense advantage of having their records before him for selection of his approach. The north face had not been attempted by six out of seven previous expeditions because it was considered the most dangerous of the routes. His party decided to accept the challenge of the north face.

One of the most dangerous and difficult obstacles on the northern face was the awe inspiring over hanging blue shadow of death - 150 feet of ice wall that kept sending down avalanches at will. No one had so far dared to cross that ice wall.

On account of the expected shortage of porters in the Badrinath area during the yatra season, he sent Mulk Raj and Chowdhury well ahead to Joshimath on the 25th of May to make the necessary arrangements. The remaining party with 5,000 pounds of baggage left Delhi on the 27th of May. The Chairman of the sponsoring Committee came to see them off at the Central Vista Mess. His final word of caution was that Bull Kumar must bring back all the members safely.

They called on General Williams at Roorkee en route to Rishikesh. He advised them to feed well and climb slowly. His parting words were, "Boys, Nilkantha is a difficult and dangerous mountain. Be careful."

They reached Joshimath on the morning of the 29th of May 1961. 'Mulki' had already arranged the mules but an important item – ladders - could not be carried on them as the track was not easily navigable. In the evening, they had a grand party in the local Officers' Mess; A fitting celebration of the eighth anniversary of the first ascent of the Everest. The team was given a very warm send off the next day.

They reached Mana village the first day. 'Chow' had rounded up about thirty porters, who, however, demanded extra wages and smaller loads. In their anxiety not to lose them, they accepted their demands. The party left Mana on the 1st of June but did not get very far. After an easy two and a half hour march on a Public Works Department (PWD) pilgrim track, the porters decided to camp beside the Vasundhara falls for the night. No amount of persuasion to move further was of any avail. On the 3rd of June, they established their Base Camp at approximately 15,000 ft after resolving the usual porter problems in which both Chow and Mulki had become adept. It was rough going over the terminal moraine of the Satopanth glacier and the last 1,000 feet climb over loose rocks and snow was exhausting. Nilkantha then stood before them. No sooner they had reached the Base Camp; they received a three-avalanche salute. Perhaps that was intended as a welcome; but they were not amused like Victoria.

Nilkantha had three distinct ridges: Southeast, West and Northeast. From Badrinath they could see the Southeast ridge and part of the Northeast ridge. Any attempt from the first seemed impossible. It rose some 8,000 feet and was swept by avalanches. Of the other two the West ridge appeared the safer. The beautiful white dome of Nilkantha, supported on ice cliffs, is separated from a lower band of black granite rock by an almost vertical 150 feet ice wall formed by glacial action.

Their conundrum was to find a route to the foot of the ice wall that would be free of avalanches, cross that, negotiate a dangerous ice shelf on to the West ridge and finally traverse the summit ridge over a distance of about 800 yards to the top. If they could get to the summit ridge, the peak was theirs. To get there would require stable snow conditions that are found only immediately before the monsoons.

It was not difficult to get to the West Col (18,000 feet) from where a very steep rock section joined the summit ridge. On the 4th of June, Chow and Bull Kumar carried out a reconnaissance of the North face towards that barrier without being bombarded by avalanches. Mulki set off the next day to establish Camp-I which is nearly 1,500 feet above the Base Camp and to prospect a site for Camp- II.

They had to decide whether to follow the West ridge or attempt the North face. Bull resolved to go to the West Col himself and see if it offered a route to the summit. On the 5th of June, Grewal, two Sherpas, a porter and Bull Kumar went up. It was a strenuous climb from 15,000 to 18,000 feet and the avalanches made it most hazardous. They kept one Sherpa and sent down the others. They were so excited about the possibility of finding a portable route on the West ridge that they climbed some way up a little peak that very evening in order to have a good look at the terrain. Bull Kumar felt they had a fair chance and, in anticipation, sent word to the rest to move up to the West col.

Sherpa Phurba and Bull started out early next morning to force the two gendarmes standing sentinel in front of their tent to Nilkantha. The view to the South of the Col was superb. Below them was a long stretch of snow with an emerald lake ensconced in snow-covered peaks. North of them, the majestic "Peak of Avalanches" was momentarily calm. A pair of black crows lent beauty to the scene. These birds are always quick to follow human beings; Grewal and he recalled the black Himalayan Caravan that trailed them on Everest right up to Camp- V.

The rocks were loose and the nearly vertical wall of about 2,000 feet that confronted them would have made anybody's knees tremble. As they climbed, a piece of stone came hurtling down towards Bull Kumar's face. A boxer's intuition brought his right hand up in defence; that deflected the stone and he escaped with a minor injury. Phurba and he kept up a good pace and crossed the "Twins" and another set of three towers that a previous British expedition had named **"Ugly Sisters".** They felt it highly improbable that at such an altitude and wilderness any association with the fairer sex should appear ugly, and accordingly rechristened these rocks as **"The Three Sisters."** Any port in a storm, so to speak! Besides, they were not exactly Cinderella's step-sisters!

A little further on, they were faced with a formidable vertical rock. They had been rock climbing for five hours then and it was impossible to go further. If that obstacle could be negotiated, there might have been a chance of finding a route to the top. However, Bull Kumar felt the situation could not be salvaged and they turned back.

On the 7th of June, they dismantled the camp and started descending. Bull Kumar saw Sharma and his party leaving the Base Camp for the West Col in accordance with the message he had sent earlier. However, since their plans were altered, he signalled him to stop at a point from where the party could be diverted to the North. He also established contact with Mulk Raj and summoned him from Camp-I for a conference. The three parties met at 10:00 am just below the West Col glacier. Bull Kumar got first hand reports from Mulk Raj who had reconnoitered the North face. After weighing the odds, he finally decided that they should attempt the mountain from the North. Sharma was sent to establish Camp-II on that face while the rest returned to Base Camp.

On the 8th of June, Chow and two Sherpas were sent to Camp-II at the height of approximately 18,000 feet. That evening the ice wall above Camp-II shattered and came thundering down. A feeling of consternation enveloped them all; if any ice were to have fallen in the vicinity of Camp-II, they would have had to change the site. No ready alternative appeared possible for such a contingency. They contacted Camp-II on the wireless and were greatly relieved to know that all was well.

The relief however, was short lived. At 5:30 pm, All India Radio announced that the monsoons would strike that area on the 11th or 12th of June. They were in dismay! Bull Kumar had selected that time of the year especially after studying the meteorological data for the past fifteen years. The monsoons usually reached that region about the 25th of June; and he had calculated on making a bid for the peak around the

20th of June. The 15th of June might have also been alright. Now they had only two days! Trust the weather to act whimsical.

There was only one Sherpa available at Base Camp. Together with him Bull Kumar started for the Camp-II after telling Chow and Sharma over the wireless to work on the steep granite spur and make for the foot of the ice wall. As he climbed, his gaze was fixed on the slope of the rock that was holding back the others. He thought the ice wall would be negotiated by the evening; thus, enabling them to camp above that and attempt the summit in the next two days. But on reaching Camp-II, he discovered that it was quite impossible to rope the last 50 feet of rock below the ice wall. It was snowing and the rock was steep and slippery. He took over and managed to gain some ten feet. Two hours later he had not got much further when his foothold suddenly gave way and he fell down the entire length of the rope. They returned to Camp-II very dejected. They were one day closer to the monsoon.

Early next morning, Chow and Nawang Tshering set to work on the ice wall. Phurba Sherpa and Bull Kumar left Camp-II a little later but progress was slow. Finally at 10:00 am they stood at the foot of that formidable barrier. The ice was sea blue and hung over them like some fantastic monster mocking its prey. They moved hurriedly towards the western end of the ice wall where they thought they might cross it more safely because the pressure of the glacier was far less on the sides than in the middle.

Bull's rope slowly took over from Chowdhary and Nawang Tshering who were cutting steps ahead of them. He located the crevasse along which he had originally thought of attempting a crossing of the ice wall as that seemed to offer an easy approach from the Base Camp but that then appeared like a death trap as a sheer wall. He therefore, decided to try a little further on. They had to keep going for some time before he could select a more favourable point. They spent four hours driving in pitons and fixing ropes but all in vain. To climb the ice wall would have required at least three twenty-foot ladders. They had some at the Base Camp but there was not time to bring them up. It was then 5 pm and the stones from the top began to rain down like bullets, whizzing past them in the cold air. When ever the rumble of stones was heard the Sherpas would dive for the nearest head cover under the rocks shouting *'pather ayo, pather ayo'* (stones falling, stones falling).

At that stage Chow came up and suggested that since they had already risked so much, they should go through to the gully on the extreme right of the ice wall. The new route suggested by Chowdhary was one they had rejected at the Base Camp as being far too dangerous. The gully was not merely vulnerable to avalanches from the entire flank of the ice wall but also drained the avalanches coming down from the dome of Nilkantha. The gully was like an inverted funnel with its mouth below them. The stem that Chow advocated if negotiated might be a little safer on account of its being narrower and at a higher altitude. Bull Kumar was against Chow's proposal; but he insisted and Sharma nodded in agreement. *"Phurba, chalega"* (Phurba, will you come), Bull Kumar asked? He turned and replied, *"Bara bara khatra dekha per aisa khatra nahin dekha. Lakin jahan sahib jaiga ham jaiga"*. (He had seen very great dangers, but none like this one. But where ever the Sahib went, he would follow).

They had no sooner started towards the **'suicide gully'** than they heard the crash of stones and ice. They put their rucksacks over their heads and waited. Fortunately, they were not in the **'hell fire line'**. The stones began to come down more rapidly and frequently but they pressed on till it became dark. Chow advised them to camp on the rocks and leave the rest to God. It was, however, decided that one of the Sherpas should carry out a recce of the area. Lhakpa Lama was roped and sent into a nearby cave; he kept going and disappeared with a hundred foot of rope behind him. Repeated calls brought no response. They tugged at the rope and finding it loose, started rolling it and all of a sudden Lhakpa surfaced with a broad smile. *"Hum bahut door gaye, camp ki jagah hai"*, he said. (I went very far and there is a place to camp). They descended into a cavern with blue ice walls and ice chandeliers under which they pitched two tents at a height of approximately 18,900 feet. It was a remarkable night. The ominous crackling from the direction of the ice wall and the constant fear of the cave collapsing kept them awake and praying all night long.

Morning came without mishap. The weather, however, was extremely inclement and Chowdhury, who knew quite a lot about meteorology, remarked that monsoon conditions had set in. Bull Kumar was in a dilemma. The alternatives were withdrawal or battle with the monsoon. Smythe had written that Nilkantha should not be attempted during the rains for two reasons - the steep slopes of the mountain made any route vulnerable to frequent avalanches; while its location at the foothills of the Himalayas offered little prospect of a break in the weather.

After considerable deliberations, Bull Kumar decided to spend two more days on the mountain in the hope that the weather might miraculously improve. Speed was essential and it was necessary to travel light. As they planned to stay up for no more than a night, Bull Kumar decided to take only a small one man bivouac tent for the seven of them - three members and four Sherpas, two days rations and kerosene oil, about fifty ice pitons and four fixed ropes. He even discarded the sleeping bags and decided instead to carry two air mattresses on which they could sit.

It was clear that it would not be possible, to climb the ice wall at the point selected earlier given the lack of time. The only alternative appeared to be to follow the "suicide gully". Bull decided to head the first rope with two Sherpas; the second, consisting of Sharma, Chow and two other Sherpas followed 50 yards behind him. They set off at seven in the morning with the weather deteriorating steadily. Visibility was zero beyond a few yards; and they hoped against hope that they might get enough of a break that would enable them to locate the route to the top. They traversed the edge of the ice wall and soon reached the "Suicide gully" which Bull then saw for the first time in its entirety. What he saw hardly relieved his anxiety. On their left the ice wall was badly broken up at that point, while to the right was a small ice spur; between the two was the gully, pock marked with scars made by the boulders carried down by the avalanches.

The gully turned out to be far steeper than they had imagined. Every step had to be cut and as they worked on the ice, he could not take his mind off the twin dangers of avalanches and the overhanging ice wall on their left. Two hours of determined effort brought them to a huge rock upon which they rested for awhile. Had those at the Base Camp seen them from down below they would surely had thought them to linger in a death trap!

Bull felt that the only course open to them then was to abandon the gully and move left on the ice wall that appeared negotiable at that point of time. Phurba took the lead and in about half an hour they were over. They felt a surge of joy and hope. He then felt the possibility of climbing the peak and returning within the two-day deadline.

At 2.00 pm, they reached the foot of a rock spur that seemed to lead up the mountain. That would have been the shortest way up but he was not sure whether it really did go right to the top. Clouds hung over the mountain and they could not get a clear view of the ice shelf on the West along which they had decided to approach the summit ridge. They waited for the clouds to lift and in the meantime climbed another 20 feet to get a better view. Fortunately, the clouds lifted momentarily and to Bull's relief, he saw a route that traversed a huge *couloir* and went up a glaciated incline to meet the summit ridge on the extreme west.

They were about to move forward when two stones avalanches came hurtling down, one after the other, across their line of advance. Sharma was hit on the thigh but escaped without serious injuries. Dusk was approaching and it was quite obvious that it would be foolish to cross the couloir at that hour of the evening. At the same time it would not have been impossible to camp where they were; nor could they return to Camp-III. Thus, they were forced to bivouac in a small crevasse about two feet wide just below the rock at an altitude of about 19,650 feet.

That was "Camp-IV". There was the disconcerting prospect of being buried under an avalanche but there was no alternative. They were packed like sardines; seven of them crushed together in a tiny one-man tent. Once inside, it was physically impossible to change position. They felt suffocated before long and in desperation punctured a hole in the tent. That was no solution as the snow soon started coming in through the opening.

Bull Kumar rose at 4 am in the morning and woke up all others as only one person could dress at a time; it took four hours to strike camp. The Sherpas were reluctant to go any further. One of them said to Bull, "*Pani nahin hai, khana nahin hai, barf padta hai aur sleeping bag bhi nahin hai. Abhi aage jana theek nahin hai*" (There was no water or food. It was snowing and there were no sleeping bags. It was unwise to continue). Bull Kumar told Chow to talk to the men in Gorkhali and tell them that having crossed the "suicide gully" and survived the night with all the odds against them, they should not give up then. The Sherpas were soon persuaded and they immediately began work on the *couloir*.

Three hours later they were on the ice shelf which was almost vertical. Bull Kumar sat to work on the face with two Sherpas. Chunks of ice flew out under the axe and those below had to shield their heads with their rucksacks as best as they could. A piece of ice hit Chowdhury on the wrist and smashed his watch. After about a hundred feet of step cutting, they discovered a snow gully that took them to the summit ridge. They sank knee deep in the snow but were glad that there were no steps to be cut. They halted at a point about 40 feet below the ridgeline.

He had planned to make a bid for the summit and return the same day but was wary of doing so since the second rope had not yet come up and the weather was extremely bad. Despite the knowledge that they had no more food, he decided to spend another night on the mountain and attempt the peak the following day. At 4 pm Sharma, Chow and two other Sherpas arrived and Bull Kumar told them of his decision and they were soon looking for a site for "Camp- V".

The wind ruled out camping on the ridge. Luckily they discovered a huge crevasse nearby. Closer inspection revealed a passage that descended to a level floor providing an excellent refuge for the night. They settled in that camp that was almost level with the first hump on the summit ridge at a height of approximately 20,450 feet. The ice formations around them bore an astonishing resemblance to human and animal figures. Dinner that night consisted of half a toffee, an inch of red chilies and a cough tablet each. They felt pangs of hunger and suffered cramps since no movement was possible in their little tent. But they had shelter from the wind and the snow.

The weather on the 13th of June was good. Two summit parties were organized - the first consisted of Chow and two Sherpas Tashi and Nawang Tshering. The second party consisted of Sharma and the two other Sherpas. Chow's party set off, but even after over two hours of gruelling effort they found it virtually impossible to make the summit ridge. They had underestimated the magnitude of the difficulty ahead. The choice then was either to get exposed to another day on the mountain or abandon the attempt.

Bull Kumar was faced with a very difficult decision. As leader it was his responsibility to ensure the safety of the team and not take any undue risk. Clearly there was no peak worth risking precious lives. The mountain was there and another attempt could always be made in more favourable conditions. On the other hand, they had taken great risks to reach a point that no one else had ever attained before. The prize seemed within grasp; but the monsoon had broken. It was snowing heavily.

They had already spent three nights of intense discomfort which had denied them sleep. They had only four pitons left and could not afford to lose them as they were indispensable for coming down the mountain. They had no food. Bull Kumar thought of abandoning the attempt but Sharma pressed him to continue and his insistence finally tipped the balance.

Sharma was bubbling with excitement. "Chow had already made a backbreaking effort and another attempt by him was quite out of the question. Bull's presence at Camp-V was imperative. Anything could happen and it was essential for the leader to be available to take whatever decisions might be necessary. He accordingly selected two Sherpas, Phurba Lobsang and Lhakpa Gualbu Lama. The latter, with their experience of many peaks and all kinds of weather, were a little apprehensive, but their innate sense of gallantry responded to Sharma's infectious enthusiasm. They joined him.

Three steep faces rising about 400 to 450 feet barred the way to the summit ridge. Chowdhury's party had already worked on that obstacle partially. Sharma and the two Sherpas then set to the task with great vigour and by 1.30 pm had surmounted that crucial hurdle, and crossed the ridge just short of the second hump. The going was not easy; it was snowing and they were slashed by occasional gusts of wind. They closely followed the ridge on the South Face of the peak.

As each hump was passed, another higher one revealed itself ahead. They continued doggedly, each encouraging the other, and at last struck the horse-shoe eminence ahead of them. That was the summit. It looked so different now; it was no longer the apex of a cone. It was fifteen minutes past five in the evening.

The three of them remained on the summit for about fifteen minutes. Sharma made an offering of flowers and garlands that he had carried up with him from the plains. He burnt some incense and planted the national tricolor and the IAF flags on the peak. The latter had been given to the expedition by the IAF Trekking Society that had attempted Nilkantha in 1959.

It was already getting dark as the proud trio began the long descent. Only then did they realize how late it already was. It would have taken them at least five hours to reach Camp-V where the rest were anxiously awaiting their return; but that now seemed out of the question. They were left with no alternative but to brave the night out in the open between two of the lesser humps on the summit ridge. Although thoroughly spent and without food or water, they braced themselves to face the menacing hours of darkness. There was a gale blowing and it snowed intermittently. They were all but frozen and counted the long, slow tortur~~ed~~some minutes of the night on their feet, hugging each other in a desperate fight against the remorseless elements of nature.

Meanwhile, the failure of the summit party to return to Camp-V filled others with grave foreboding. Thoughts of death defying possibilities crowded their minds rendered all the more fearful by the hostile environs. They waited up until about ten o' clock in the evening, flashing torches into the night every now and then to direct their companions to the camp. But there was no answering sign or sound to suggest that the summit team was returning.

Their companions at the Base Camp and Camp-II were also deeply worried. They had not been able to see them due to bad weather and their fears could be judged from the fact that they drafted the following signal for dispatch to the sponsoring Committee: "The leader, Flight Lieutenant Chowdhury, OP Sharma and four Sherpas have been missing since the twelfth."

At about 5:00 am the next morning Sharma, Phurba and Lakhpa Lama resumed their epic descent. At one point the three slipped and fell on the hard morning ice. One of the Sherpas lost his ice axe. A little later, while repelling down, a piece of ice broke off and fell on Sharma's head making a gash a quarter inch deep and three inches across. The three staggered into Camp-V at about 10 am accomplishing the mission. Whole team was overjoyed to see them and no less delighted to learn of their splendid success.

Bull thought it would take them only a day to descend to Camp-II but such was the weather and the terrain that it took them three days – 82 long hours without food or water! They made to Camp-IV by 6:00 pm on the 14th of June. It was snowing. He pressed himself into a corner of their one man tent. The tent zip tore open during the night and his feet got exposed to bitter cold. He requested the Sherpa who was lying over his legs to get up for a moment so that he could pull in his feet. But they were so tightly packed that he could not budge an inch. By morning Bull's feet were badly frost bitten.

They did not follow the "Suicide Gully" from Camp-IV to Camp-III but came straight to the edge of the ice wall and decided to rappel down. That was not easy. Their fingers were so numb that they had to hold the rope with their teeth. However, they crossed over without any mishap. Camp-III that they reached on the 15th of June was a home away from home. They were given some khichree and orange squash that had been sent up from below and they gradually began to feel that their bodies and souls might at last again be able to perhaps hold together.

They had only one piton that they fixed to the ice-wall. Since that was not enough, they secured an ice-axe in a nearby crack and tied the rope to that also. Having tested both ends, they lowered Tashi who went down without much difficulty. He was to fetch a piton from Camp-III and when they had procured that, they finally secured the rope at three joints. That was quite adequate. As the space on the platform was very restricted they called for one volunteer who was to rappel down immediately. Nawang came forth. The carabiner was tied to his sling and the rappelling rope was wounded around that twice. Off went Nawang without delay. After sliding down six feet or so, he tried to control his rappelling by holding the rope in his hands. But his fingers were numb and he could not grasp the rope to arrest his motion. He took a quick terrifiying look at the depths to that he was sliding and then cried out like a child, *"Kumar Sahib! Bara Sahib, bachao!"* (Kumar Sir! Big boss, save me!). He continued to slide rapidly and helplessly, possibly to his death.

Bull called him out, "Nawang, hold the rope fast!" But he kept shouting "*bachao, bachao!*" (Save me, save me!). Nawang very bravely made a last effort to save him. He bit hard into the rope and his strong teeth held it. He hung there speechless and helpless. There was a temporary relief, for he was safe, but Bull was afraid that his teeth might break with the weight of his body or that he might not be able to hold on too long because of the extreme cold. A shudder chilled Bull's spine. There was no other rope available. Chowdhary, Sharma and Lhakpa quickly undid their rope and they secured it to the pitons. Bull's fingers were also numb. But Phurba quickly tied the rope. Someone had to go down immediately to Nawang. But who would be that?

Bull's feet, from the ankles down, were like hard, gnarled wood. He had been sick with dehydration the whole day. Phurba had done a lot for them. Bull looked at him in silence as though to ask if he would go down to help Nawang. He had earlier seen him coming down the rappelling rope but then he had always used it as a fixed rope.

Could Sharma go? Bull knew that Sharmas was suffering from greater dehydration than the rest of them. In any case he was not adept at rappelling. Whatever was to be done had to be done instantly. While he had been hanging, Nawang had drifted to the steeper face of the ice-wall. If he slipped accidently, no one could have saved him and Bull would never have forgiven himself.

Quickly he prepared to go after Nawang himself. He kept hitting his bare hands mercilessly against the ice-axe, the rope and his boots in turn, to restore some blood circulation. He hopped and jumped on his feet too but it was of little use. Everything below his ankles was dead to all sensation. The swelling had come up to his calves. He put his carabiner in the second fixed rope and lowered himself.

His mind was confused. He did not know how he was going to save Nawang. He could not hold the rope with his hands. So he put the upper end of the rappelling rope between his teeth and twined the lower end around his arm. He lowered himself slowly, wondering all the while what he was going to do. The rope was cutting into his arms. He was also afraid of losing his teeth. Soon he reached upto Nawang. There was a deathly shadow on his pale face. He had obviously lost all hope of life. The fear of death had contorted his face into an ugly mask. He tried to smile but in vain.

Bull held the lower end of his rope and threw it up. It did not reach Phurba. Once again He tried and missed. Both the times the end of the rope came down and slapped against their faces. The rope had not reached half way up even though the distance was not more than 20 feet. Bull was getting extremely desperate. Finally, he coiled the rope like a cowboy lasso and threw it up with all his dwindling might and he prayed soulfully. Luckily that time Phurba caught it and thus, Nawang was safe.

The next step was to join the two ropes together. Having learnt from the bitter experience, they did not lower the loose end of the rope and kept it with them. That way they lowered themselves one by one.

They descended to Camp-II on the 16th of June and to Camp-I the following day. Bull was immediately evacuated to the Base Camp on a stretcher. Mulk Raj and Grewal had made all the arrangements for winding up the Base Camp and half of their equipment had already been sent down to Mana.

As they left Nilkantha, Bull lifted his head from the stretcher and looked back at the mountain lion, where they had spent the last few days with mixed feelings of hope and despair. The whole sequence of events passed through his mind and he lived them again. He thought of the splendid teamwork and comradeship of all the members of the expedition including the Sherpas. He recalled the ice wall and the five unforgettable longest days they spent beyond that fearful barrier, clinging to each other through the long, sleepless, weary and bitterly cold nights.

The throbbing pain in Kumar's feet woke him from his reverie. The other members of the expedition limped along, barely able to walk on their frostbitten feet. They were weak but undaunted. He was deeply moved and thankful to God Almighty for the success as they had not paid too big a price for that great and unique pilgrimage!

At 4:00 am Bull realized that he had been frost bitten - it was a terrible thought but it was unfortunately true. After a nightmarish descent to Camp-II, he was carried on a stretcher all the way back. But the damage seemed to have been done and that resulted in his spending a year in the hospital with the doctors struggling to save as much as they could of his toes. They succeeded a great deal but perhaps as the leader he had to make some sacrifice to that great mountain in compliance of that onerous responsibility as an act of God's command.

While he was in the hospital, the World's first cosmonaut Yuri Gagarin was in India. The Prime Minister Pandit Jawaharlal Nehru threw a formal dinner party for him and Bull was given enormous importance and honour to sit next to him at the function. Mrs. Indira Gandhi had come to the hospital and had personally invited him.

He received honour and place of pride not because of the ascent of Nilkantha, but Pandit ji felt that Bull got the frostbite because he was delayed due to the Barahoti Mission that was the first move of his "Forward Policy" to check Chinese ingress into Indian Territory. Mrs. Indira Gandhi took a personal interest in him; and at one of the Investiture ceremonies where he was awarded the Kirti Chakra, she had asked Rajiv Gandhi to look after him.

NANDA DEVI - 1964

Life seemed dull with no prospect of climbing again and soldiering from the front echelon to the ones in the rear only, as Bull was placed in permanent low medical category, not to be posted above 7,000 feet altitude. After the failure of the Second Indian Everest Expedition in 1962, the Indian Mountaineering Foundation booked the mountain again for 1965. In preparation for the expedition, 3 more pre-expeditions were launched and Gurdial Singh was supposed to lead an expedition to Nanda Devi – then the highest mountain of India and unclimbed by any Indian. MS Kohli was to lead the Nanda Devi East Expedition and the Panch Chuli Expedition was to be under the leadership of Flight Lieutenant AK Chowdhury.

It was early May 1964, Bull was in the Army Hospital for minor scraping of dry gangrene formed on his frost bitten toe, when, two senior bureaucrats, Shri SS Khera - the Cabinet Secretary to the Government of India and Shri HC Sarin - the Defence Secretary, paid him a visit in the hospital. He was surprised how the wrath of such senior government officials could fall directly on a Major.

After pleasantries Shri Khera spoke to him, "We want you to lead Nanda Devi Expedition- can you?" Bull inquired, "What happened to Gurdial who had been nominated as the leader?" He was told that Gurdial on personal reasons had opted out. He explained to them that as per medical advice and instructions he was not supposed to climb above 7,000 ft.

"We want you to go only up to the Base Camp and stay there and just supervise the expedition". They also promised a helicopter to take him to the Base Camp. He had to make a big decision. Was it fair for an invalid climber to lead such a major expedition? On the other hand he was getting a lifetime chance to get back to his favourite sport 'mountaineering'. He hesitated initially but finally accepted the offer.

On the 20th of May, the helicopter piloted by Flight Lieutnant Chawdhury and Pilot Officer Master, arrived at Joshimath from Gauchar air strip. It was quite amusing to see both the pilots in their Pyjama suits. Their camp at Gauchar had been uprooted by the hail storm and all their belongings had been blown off. Though the pilots had spent sleepless nights inside a wet and cold Army truck they were right on time.

The flight in a helicopter through the Rishi Gorge into the sanctuary was the most exhilarating experience that any mountaineer could ever dream for. It was truly a dream flight. While flying, one could get the correct perspective of the sheer mass and grandeur of the mountains. The mountain views from the ground level got shortened and even very steep slopes look flat. The first big mountain one saw was Dunagiri, that had been attempted as early as 1883 by WW Graham; and Eric Shipton attempted it in 1936. Then there was Changbang the granite polished spire that looked so imposing and spectacular that larger mountains around got eclipsed.

On the right side of the icy scoop of Bethanrtoli stood out Trishul and behind that everything looked unimpressive. The same mountain (Trishul) seen from the south was perhaps one of the most beautiful sights in the world and to Hindus that signifies 'the divine trident of Lord Shiva'. However, as one got into the sanctuary one's true romance with Nanda Devi began. There was something special about that mountain- guarded by Trishul (Trident) in the south, having the temple bell of Nanda Ghunti (bell) and Nanda Khat (bed) nearby considered by every Hindu as the 'concert of Lord Shiva'.

There are so many stories about that mystic mountain that one could write a book on the same. Bull's only regret was that it was too short a flight perhaps 15 minutes that other wise would have taken him

10 days. When they reached the sanctuary, it was not difficult to spot the helipad made by the advance party as the smoke candle could be seen from far away.

By the 23rd of May all the stores had been dumped in the Sanctuary Camp and thanks to the helicopter, they were also able to take about 1,000 fresh eggs and 500 lbs of fresh vegetables - a sheer luxury in the mountains- especially in Nanda Devi Sanctuary. Due to shortage of porters, the Sherpas and the team members had to help ferry loads to Base Camp that was established on the 24th of May.

On the 25th of May, Nawang Gombu and Chowdhury tried to establish Camp-I but were beaten back by inclement weather conditions and were stranded in snowstorm for 3 hours. Camp-I was established the next day at the height of 19,000 ft. There was only one way to climb that mountain - from the South side, and that was through the South East Ridge. It had been referred to as 'Cox Comb Ridge' by Tilman. That ridge was so steep, narrow and thin that one can never lose the way, but there are very few camp sites on it and many a time the ice had to be cut to make platform for even the smallest tents available.

Bahuguna's party consisting of BP Singh and a Sherpa relieved Nawang Gombu to establish Camp-II. On the 30th of May Major Bahuguna left Camp-I to traverse the Cox Comb Ridge along with BP Singh and two Sherpas. As BP Singh was not fully acclimatized and found the going tough, he returned to the Base Camp. The route from Camp-I to Camp-II lay on an extremely steep and narrow ridge. On the left there was a steep precipice and on the right a sheer fall. The going was made tougher by knee-deep snow. Bahuguna made most of the route to Camp-II and the task was finally completed by CP Vohra on the 1st of June, at the height of 20,000 ft.

The route from Camp-II to Camp-III lay on a knife edged ice cum snow *arête*. On that portion a fixed rope had to be placed, to make the going safer for the porters. Camp-III was established by the party consisting of Gombu and BP Singh and two Sherpas on the 6th of June at the height of 21,200 ft.

Some of their Sherpas contrary to Bull's expectations suffered from altitude sickness. Perhaps the Sherpas of Darjeeling were becoming too urbanized. That affected their logistics build up to stock up Camp-III for further move. Bull employed the Garhwali porters at Camp-I to carry the loads to Camp-III, but they also suffered from altitude sickness at Camp-I and had to return to the Base Camp. Their essential load carrying capacity had dwindled down considerably. In desperation, Kumar tried the relay system, in that some porters carried loads from Base Camp to Camp-I. They were accompanied by equal number of porters who went empty handed to Camp-I and took over the loads from the first group and ferried them to Camp-II and returned empty handed to Base Camp. That system worked well.

He tried the similar method for Camp-III. In that case one group carried the loads and two groups went empty handed till Camp-I. From Camp-I the loads carrying group came back and the second string carried the loads to Camp-II. They were in turn relieved by third group that carried the loads to Camp-III. It was indeed a great feat by the porters. In just one day they had transferred loads from 16,000 ft to 21,200 ft where Camp-III was established. That methodology solved their logistical problems of the porters camping higher up.

Bull had been so frightened by the doctors regarding effect of cold on his frostbitten toes that he landed at Base Camp fully clad as if he was going to climb the Summit of the Mount Everest. He was wearing a down suit, huge gloves and reindeer boots and that too at the height of only 14,000 ft. However, slowly he regained his confidence and started shedding the layers. He just could not resist the urge of climbing and went up to Camp-III at 21,000 ft, when the weather was good. That was not only great for his self assurance; it boosted the morale of the others too.

Gombu and Bahuguna established Camp-IV on the 9th of June morning at the same site where the late Major Jayal had his last Camp; they found remnants of late Major Jayal's Camp, including a tin of Swiss cheese that was passed fit for the consumption by their doctor. The rock face between Camp-III and Camp-IV looked almost impossible from below. It looked like a polished face devoid of any hand or foot holds. But in actuality it did provide good footholds and handholds, and the rock was so soft that it could be climbed with crampons on.

Then the plan conceived was that Vohra's party already on their way up, would join Gombu's party at Camp-IV on the 11th of June, establish Camp-V on the 12th of June and reach the summit on 13th. But unfortunately, the weather became extremely hostile, and both the parties had to return to Base Camp. In fact, Vohra's party was caught in a vicious blizzard and coming down along the thin, narrow and steep ridge was extremely dangerous. Once Gairola was literally lifted by the strong gust of wind while heading for Nanda Devi Glacier and Vohra was able to arrest his fall. The visibility had become so poor with the blizzard that Vohra's party was forced to remove their goggles. It was a question of either saving lives or eyes. When the party reached Base Camp, all of them were suffering from acute snow blindness. Fortunately, snow blindness was not a permanent ailment and got cured within 2-3 days; but it was very painful.

The weather grew from bad to worse; and there some climbers were doubtful that they would ever succeed in climbing the mountain. Another team led by MS Kohli was climbing Nanda Devi East from outside the sanctuary when they decided to call off the expedition due to bad weather. Bull decided to stick on and went on half rations, with strict instructions not to do any exercise just to save energy. The weather cleared and improved; they could not believe their luck.

Bahuguna had developed a blind spot in his right eye at Camp-IV and was advised by doctor to do no further climbing. It was a great set back to the expedition because he was one of their summiteers.

On the 16th of June, the first support party consisting of BP Singh, 3 Sherpas and 2 porters was sent up. Apart from carrying loads to Camp-IV for the Summit Party, that party was also to make route that had been obliterated due to continuous snowfall for the past six days. The Summit Party consisting of Vohra, Gombu and Dawa Norbu had left the Base Camp on the 17th morning and went straight to Camp-III.

On the 18th of June, they at Base Camp saw eight ant like figures climbing up the rock face at Camp-III. Everything seemed to be moving according to plan. On the 19th of June, the Summit Party with

Sherpas Sona and Nima, in support, moved up to Camp-V. The already tough going around the steep rock ridge, was made awfully difficult at places due to fresh snow. After about 3 hours of steady climb the party reached a huge wall of white rock and above that Camp-V was established at the height of 24,500 ft-the highest camp put on that mountain ever! Higher even than Nanda Devi East.

Butane Gas carried for Camp-V accidentally fell down the ridge and hurled 8,000 ft below. Luckily the party was carrying the emergency, one man cooker borrowed from the Army that worked with solid fuels. It came in very handy even though it was not enough to make sufficient liquids for the three of them. On the 20th of June, the Summit party consisting of CP Vohra and Dawa Norbu left the Camp at 7.15 am in excellent weather conditions. The team felt fit even though they had not slept well the previous night. The going was good, but the altitude was telling on the climbers. The party reached the yellow rock, the point from which Major Jayal's party had returned in 1957.

CP Vohra, who was suffering from severe pharyingitis, realized he was delaying the progress of the entire party and might jeopardise the chances of others going to the top. That occurred to him when he remembered that Tilman had taken 9 hours to reach the summit from the last camp. Vohra dropped out 500 ft below the summit and insisted that Gombu and Dawa Norbu carry on. Gombu was reluctant, but was forced by Vohra to attempt the peak.

After leaving Vohra, the party climbed about 400 ft when they found two snow regions, one on the left and another on the right. Gombu climbed the left ridge that took him to the base of white rock that encircled the ice and snow plateau of the summit. That rocky portion was steep and loose. After crossing it, the Summit party cramped on 100 ft of hard snow before they stepped on the Summit of Nanda Devi at 11.30 am. The party stayed there about half an hour and returned leaving a national flag hoisted on an ice axe.

On return to the Base Camp Gombu told Bull that his camera had not worked on the summit as the shutters froze due to cold winds. Bull's heart sank into his boots! Who would believe them? Mountaineering in India was in its infancy and anything momentous accomplished by any Indian mountaineer was looked upon with suspicion unless there was hard core evidence.

Nanda Devi already had a reputation of sorts - Maj Nandu Jayal's party that was considered one of the toughest and had Sherpas who had gone to last camp on the Everest had failed; Gurdial and Hari Dang had made attempts. There was an added negative factor; only a week ago Captain MS Kohli had called off the expedition due to bad weather. Every one would question, how the weather could be good for one party and bad for another on the same mountain!

Bull had already suffered the consequences of such a situation earlier, as leader of Nilkantha Expedition in 1961. He had sent a strong team for the summit attempt. While reloading the camera the film of the summit photography had been exposed accidentally. They did not have photographic proof. A huge controversy followed as some rich Bombayites chairborne mountaineers questioned the climb. On his request Indian Mountaineering Foundation constituted an inquiry committee that upheld the claim. However, the damage had been done. His enemies twisted the facts and conveyed a message to the world as if Kumar had made a wrong claim.

He was again in a precarious situation. On one hand he had to trust his climbers, while on the hand there was no proof. After a lot of thinking he decided to shoulder the responsibility and sent back the message that the Nanda Devi had been climbed. But all the time he felt uneasy and insecure. The last thing he wanted to repeat was the Nilkantha controversy again. After couple of sleepless nights he requested the Defence Secretary Shri HC Sarin for aerial photography. His request was accepted and the results proved his judgment was right. The second ascent of Nanda Devi after 28 years had been successfully attempted by the Indian climbers.

EVEREST AGAIN – 1965

With the success of the Nanda Devi Expedition, Bull Kumar was made the Deputy Leader in the 1965 Everest Expedition under the leadership of his old friend Commander MS Kohli. He felt then, that Everest should be within his grasp.

They set up their headquarters in Delhi. With only 5 and half months to plan they had to move quickly with the planning and team selection. They divided the work among themselves. While the leader made a list of equipment and handled all the formalities, he was dispatched for Kathmandu to arrange for the Sherpas. Nawang Gombu proved to be of great help, he even travelled to Namche Bazaar to pay and talk to the Sherpas himself.

By the time Bull returned to Delhi, Kohli had succeeded in arranging for the funds. They spent four days over the equipment list. That time, also most of their items were to be bought or manufactured within India and only some equipment needed importing. Like the oxygen masks that had proven to be a hassle at the summit in 1960, the Americans had created a new mask that was more durable at those altitudes. They had to import that particular brand or manufacture similar one in the country.

As a Deputy Leader, Bull's responsibilities were far more than in 1960 expedition. The order of business was negotiated with the foreign firms and Ordnance Factories persuaded to complete the work on time. Meanwhile, Kohli had to head towards Darjeeling where the HMI was holding another Pre-Everest Course for the team selection and he was to administer the expedition on Rathong-one of Kanchenjunga's smaller peaks. Kohli's mind was clear on the selection, nothing but the best would suffice. And that was exactly what he picked.

He returned to Delhi with a broad smile on his face, he was satisfied with the list of climbers he had chosen for the expedition. They were Nawang Gombu, Gurdial Singh, Sonam Gyatso, CP Vohra, Major Mulk Raj, Sonam Wangyal, Captain AS Cheema, Captain HPS Ahluwalia, Captain HV Bahuguna, Captain JC Joshi, Captain Soares (doctor), Dr. DV Telang (doctor) and GS Bhangu (wireless operator).

It was a well balanced team. There were ten married and nine bachelor members. Nine of the nineteen had been to the Everest before. The oldest member, Sonam Gyatso, was 42 years old while the youngest, Sonam Wangyal, was only 23 years old. The average age of the expedition was 31 years. That was the best possible team they could have amassed, rich in experience and young in years.

With the team selected, their headquarters were busier than a beehive. All hands were helping, optimizing team spirit and group cohesiveness. They had a deadline and they managed to work together to get every little detail down. It was decided to stick to the old route. Bull was keen on trying the Western Ridge route but the Sponsoring Committee would not have that high a risk on their hands. Thus, they took the same route as they did in 1960. From Jayanagar on 26 February to the Base Camp (17,800 ft) on 22 March, paying their respects at the Thyangboche Monastery on the way; for Bull it was a feeling of déjà-vu. The river, the monastery and the villages spread across the Sherpa land, all felt like old friends.

Even the Khumbu ice fall seemed friendly to Bull. That ever-changing jungle of ice was a formidable obstacle. The glittering, smooth towers threatened to tumble down with thunderous rage; huge unshapely blocks rumbled, screeched and came shattering down at will. Frightening crevasses yawned, gulping tons of debris hungrily and closing in as fast. His old friend, Everest, was at its tricks again. This time they were determined to win.

For three days the teams worked on the ice fall. When the 27th of March dawned, they decided to have a little gamble. Bull said the team would reach Camp-I that evening but some said they would not reach till the day after. They watched them ascend with their binoculars trained on the mighty fall. When the first tent was erected and their doubts put to rest over the wireless, the whole camp rejoiced. Even those who lost the bet were happy to do so. Across the ice fall in 4 days, that was the quickest ascent to the Western Cwm in the history of mountaineering on the Everest!

Soon the route was cleared and declared safe for porters. The members and Sherpas started to ferry loads to the Camp above the icefall. Although each trip to that tottering chaos meant a real hazard and its successful completion a great relief, loads were carried willingly and cheerfully. Soon ferrying loads up the ice fall became a routine and its hazards were forgotten. Sherpas and members carrying 50 lbs each on their backs were moving up and down enthusiastically, singing songs, returning to the base in the afternoon in long strides.

While the loads were ferried, teams continued to open the route to Camp-II or the Advance Base Camp. Crossing the Cwm seemed simple at first but Bull knew first hand that the simple path hid its fangs beneath. The members proceeded with caution, probing the ice ahead with their axes before proceeding. They managed to avoid and bridge many hidden crevasses. Finally, they decided to establish the camp in a gentle hollow, beyond that lay a long placid ice-slope at 21,300 ft. The camp's location in the depression hid it from view. They had to hoist a large flag over the highest tent to make it easier to find.

For many that camp was just an enroute stop over while others made it their home. Lala Telang, for example, thoroughly enjoyed his stay there; one could hear his booming voice from a distance. Though the same could not be said for Danu the assistant cook who lost his voice on reaching Camp-II and did not regain that until later, when the warm liquor at Thyangboche restored his vocal chords.

Camp-III was established at the foot of the Lhotse face by Gombu and Ang Kami on the 7th of April. Meanwhile, they had another problem growing at the Advance Base Camp. The food rations were falling short. There was an ever increasing demand for fruit juices and the supply just could not meet the demand. Their quarter master,

Major Mulk Raj, was perplexed. He had planned everything according to a balanced diet for the climbers and he was sure there was not a mistake in his calculations of stock. Soon they found the cause of the problem. The choughs and the goraks, mountain birds with strong beaks, had been sneaking in their food storage. Their powerful pecks were able to puncture holes in the compo food boxes.

Work continued up the Lhotse to reach the South Col. Gombu and Cheema, Sonam Gyatso and Sonam Wangyal, working in pairs, spent three hectic days puffing oxygen at two liters per minute. They finally, succeeded in clearing the route on the 12th of April and the team established Camp-IV at 25,000 ft. They fixed rope for hundreds of feet across the tortuous reaches of Lhotse till the Yellow Band. They found about 300 feet of rope abandoned by the 1963 American Expedition and utilized it as a hand-rail across the Yellow Band.

As the deputy leader, Kumar was handling the logistics and managing the ferrying of loads from the Base Camp. Since he was posted at the base and the higher parties were sent there to rest on a routine basis, it gave him the perfect opportunity to talk to the members and judge their levels of fitness and mental state. Lieutenant Cheema was an old friend and both of them had climbed Barahoti together in 1961 and he was the toughest man in that expedition. Actually, he was included in that expedition on his recommendation even when he had not done the mountaineering course. He was fit as a fiddle, and his work on the mountain was exceedingly commendable. He was the perfect choice for a place in the summit party.

Kohli climbed to the higher camps to supervise things. The good weather held on and their first ferry reached the South Col on the 16th of April. Two days later, another ferry of 16 repeated that performance. As usual, the South Col ferries rummaged around the 'highest junkyard in the world'; and the luckier ones returned with hundreds of feet of cine film left by the Americans, as also oxygen regulators, strips of tent fabrics (that they used as scarves), and most surprising of all, Hari Dang's wallet containing a couple of hundred rupee notes in the Nepalese and the Indian currencies of the 1962 Indian Expedition.

When Vohra was at the Base, they had a chance to converse. He was a geologist who went on to retire as the head of the Indian Geological Service. Throughout the expedition, he kept adding pieces of rocks to his rucksack for tests and research later. There came a point when he had collected over 10 kg of rocks and asked Bull to share his burden. The man ferried loads, led parties up the mountain and added heavy rocks to his collection while still managing to remain in perfect shape. He also was a good candidate for the summit.

The morning of the 20th of April dawned bright and clear and their first summit party consisting of two pairs—Gombu and Cheema, Gyatso and Wangyal —supported by Gurdial and Kohli moved up to the Advance Base Camp. A team of 14 strong and selected Sherpas accompanied them. It was the first time in the history of the Everest that a summit attempt was being launched so early in the season and, if all went well, the summiteers might reach the top by the 27th of April. The weather seemed fine and the Lhotse face, usually wind-swept with frenzied gusts of driving snow, now looked serene and peaceful. There was no plume on the Everest and their hopes were sky high.

The special weather bulletin broadcast by All India Radio, however, indicated bad weather over Everest on the 25th of April, owing to low pressure area in the region. They considered it wise to stay put till the forecast was more favourable. True to the weather forecast, the gusty winds swept the camp on the morning of the 25th of April. The tempest surrounded the entire mountain. The stillness and calm, that they had hitherto enjoyed, was now a thing of the past.

On the eve of the 26th of April, favourable and settled weather with moderate winds was forecast from April the 27th till the 1st of May. Waking up to a beautiful morning on the 27th of April, the first party along with Gurdial and Kohli, proceeded directly to Camp-IV from the Advanced Base Camp and the following day they moved to South Col. It was then that the winds picked up momentum. They spent over two hours pitching tents in the strong gale. The moment the tents were hospitable, the climbers stuck in their respective sleeping-bags praying the winds would calm down. But alas, the Everest had other plans! The winds continued to blow wildly and after spending two nights in the harsh weather the dejected party returned. Everyone descended to the Base Camp. Advanced Base Camp was also deserted, except for Danu and a Sherpa who were left there to man the camp and give them hourly reports on the weather.

Kohli let the headquarters know that they had failed in reaching the summit for a third time. He went over their logistics once more- they had the rations, the equipment and a fit team. All they needed was some good weather and they could do it. Bull convinced Kohli that the expedition was not over, as per his logistics and calculations they could still place around 11 members on the summit. Kohli called it Bull's gambler's instinct, but as the deputy leader and the head of logistics Bull had the numbers backing him on the logistics support available.

Thus they waited and prayed. The Base Camp was abuzz with activity. Everyone was trying to keep himself busy with playing cards, going for walks and exploring the surrounding regions. Food was ferried by the porters from Namche Bazaar, most of them kept their attentions on the radio, listening for the first sign of the weather clearing up.

Bull took that opportunity to accompany HPS Ahluwalia for a hike in the area and perhaps some conversation in the process. Though unfortunately for him, he was having an off day and did not seem as fit as he would expect a summiteer to be. So, Bull recommended his name for the fifth summit party. Since he was expecting a much higher order, the news was not received well by him. In fact, Bull thought HPS was holding a grudge against him. He found out later that HPS thought it was Bull's personal dislike for him that did not let Bull place him in the first summit party.

One afternoon a party of Sherpas, while moving up on the icefall, saw seven Chinese soldiers at Lho-La, a pass on the Tibet-Nepal border barely a mile from their Base Camp, dressed in drab blue uniforms. These men observed the Indian camp on the West Shoulder of the Everest before leaving the area.

The news brought some excitement into their camp. Being in the camp for so long was getting on to him and therefore, Bull sought the leader's permission to ascend up to Lho-La. Thankfully, Kohli liked the idea. Lieutenant Rana, the Nepalese Liaison Officer and Bull left base camp early in the morning with two Sherpas- Pemba Sunder and Lopsong.

In an hour they reached the avalanche zone and commenced their climb up the rock face, fixing rope to traverse along the couloirs. But when they reached the foot of what looked like a big ice-tongue descending from the pass, they found an impassable crevasse thast forced them to retrace their steps. They climbed the steep couloirs that were exposed to the avalanches before they reached Lho-La.

But when Bull looked up the Western Shoulder, his jaw dropped. There was an over-hanging glacier right above them. But there was no turning back and they decided to climb the remaining 1,000 ft. as fast as they could. Finally, at 12:30 pm, they were at the pass. It was a huge expanse of land, as big as a polo ground and almost level. It dropped down gradually to the Rongbuk glacier; one could easily ski down that route in 10 minutes. They spent two hours at the pass, looking for some sign of the Chinese. There was none. His guess was they had come to check if the Indians were attempting the western ridge; that lay on the Tibetan side of the Everest and would require additional clearances and paper work if Indians were to use that route.

They returned to the Base Camp in a tired heap, it felt good to climb after so many days in the tent. But what felt better was the reception they received. Apparently, their departure was followed by an avalanche on that route and the team thought they had perished. They all came to receive them together, Bull almost felt like they were summiteers returning to base.

The winds kept them cooped up for over three weeks. The 14th of May dawned to a still morning and Bull felt unused to so much silence after the prolonged howling of the winds. The radio reported a forecast of fair weather and suddenly in the camp was a frenzy of activity. The time to move up had finally arrived.

The summit parties were announced on the 15th. They were to start moving up on the 16th. Gombu and Cheema, as the first summit pair, started packing their gear. Sonam Gyatso and Sonam Wangyal, the second summit pair, held important discussions with 'Brig Thondup', their cook, about menus suitable for their highest climb. CP Vohra and Ang Kami were to be the third summit pair, Rawat and Bahuguna the fourth and BP Singh with Ahluwalia the fifth.

The 16th of May was a beautiful morning, clear and still - a good omen indeed! Cheema and Gombu moved to the higher camps through the ice fall. Every day a pair was to leave the Base Camp, thus keeping enough gaps for the preceding pair to move up or retreat. Kohli set up shop at the Advanced Base Camp since it had wireless contact with all the camps. Bull managed things from the base. Next day, the Sonams moved up. Prayer flags were fluttering everywhere. Sonam Gyatso's prayer wheel was majestically turning round on the red medical tent, telling them that all would be well.

Gombu and Cheema, with their caravan of Sherpas, moved from Advanced Base Camp to Camp-IV and going further to reach the South Col on the 18th of May. The winds were calmer but the South Col was as inhospitable as it was in April. Oxygen pressures were checked, food was cooked for the next day, and the summit party snuggled into sleeping-bags for the night. Cheema lay awake for some time, thinking of the unknown before him. Gombu was now an old hand at the game. He had been beyond the South Col and knew every inch of the way. He went over the anatomy of the summit ridge mentally. They were in excellent spirits.

The next morning was calm, clear and bright. Taking advantage of the weather, they left early in the morning. There was loose powdery snow in the couloirs above the South Col, and the summit party sank knee-deep. By and large, the going was good. The South Summit was visible. So was the hump below that.

Their ascent continued till they were right on top of the hump. A little below that, taking shelter from the winds, ten men went into action. After 90 minutes of huffing and puffing, a red two-man draw-tite tent was erected. Thus, the final camp was established at 27,930 ft, the highest ever camp on the Everest. They received the news at the Base Camp with great relief and joy.

An excited Cheema woke up as early as 3:00 am on the 20th of May. He roused Gombu from his sleep and they had some hot coffee before donning their gear. Then they prayed. Gombu wondered if he would do it again. That was Cheema's first experience above 26,000 ft and he was holding pretty well so far and prayed the luck would hold his mission. Then taking two oxygen bottles each, they reported to the Advanced Base Camp and left at 5.00 am. Excitement crept through the camps like a shiver down the mountain's spine. None of them had slept well. Bull was eagerly waiting for the morning, the first rays of sunlight found him outside with his binoculars trained on the summit ridge.

The day was partially cloudy, with a strong wind. The ridge was blanketed by soft snow, Gombu and Cheema went along, stamping their steps. At 7:30 am, they were directly below the South Summit. After a sip of coffee, they dumped their partially used oxygen bottles and continued forward. At 8:10 am, they were at the South Summit, where they were seen from various 'observation posts (OPs) the team had established.

Kumar had established his OP on the Pumori Ridge. The other OP's were at Camp- IV and at the Advance Base Camp. The summit party was moving rapidly and, without difficulty, they reached the well known Hillary Chimney and were seen above that after cutting a few steps and before them was the final summit ridge. Hearts thumping, not so much with exhaustion, they climbed up foot by foot on the last lap to success till they saw the American flag pole pitched by James Whittaker and Nawang Gombu on the 1st of May, 1963. Ten feet below the top they stopped and undid their rucksacks. They took out the cameras and the various flags they had carried and then climbed together. The Indian Tricolour was on the top of the world at 9:30 am on that May morning! Indeed what a marvellous achievement that was!

Cheema became the first Indian born to reach the Summit and Gombu the first one to reach it twice. Both had a tremendous sense of relief on having got a job well done. Otherwise, there were no particular emotions. It was so difficult to think or feel anything at 29,000 ft. when one removed one's oxygen mask.

They stayed on the top for about 30 minutes. The view to the south and east was obscured by the cloud. But to the north, Rongbuk Glacier and Tibet were distinctly visible. Cheema planted some silver coins given to him by his mother. Gombu left a scarf given by his wife and a statue of Lord Buddha given by his famous uncle, Tenzing Norgay.

From Bull's observation station, Rana observed their ascent through his binoculars. He exclaimed, "I can see the two of them now almost on the top." The clouds blocked their view, but a little after 10:00 am, they saw them descending. Bull immediately contacted Kohli on the walkie-talkie. Needless to say, the congratulations and excitement were in order. They had done that and the months of hard work had finally paid off. It was on the 25th of May 1960, when Bull had failed to reach the summit; and five years later that dream was fulfilled.

Cheema and Gombu confirmed the news at 1:00 pm, when they spoke from Camp-VI to Gurdial at Camp-IV. And the news was relayed to the lower camps. A message was flashed to Kathmandu and then New Delhi reporting the achievement to the Sponsoring Committee.

The weather turned bad that night. It snowed heavily over the Everest region. In fact, it snowed so heavily that the big, blue mess tent in the Base Camp collapsed under the weight of snow, pinning Lala and Bull under the debris. They had to cut their way out of the fabric.

Next Day, the second summit party - the two Sonams - occupied Camp-VI and spent a restless night. Ang Dawa had suffered frostbite in three of his fingers. Gunden was snow-blind. Da Norbu was terribly exhausted. Sonam himself had suffered a burn on his wrist (later diagnosed as frostbite). Of the two air-mattresses in Camp-VI one was punctured and useless. Wangyal slept on it. Sonam tossed in his sleeping bag, due to acute pain.

They left Camp-VI at 6:45 am for the summit. The wind was blowing hard; the snow in the air clouded their vision. Clouds were gathering. But the duo carried on and after hours of strenuous wading through the snow they were awarded. At 12:30 pm on the 22nd of May, they stood atop the Everest and spent fifty unforgettable moments there. They hoisted the flags. Sonam Gyatso placed a scarf, a statue each of Lord Krishna and Lord Buddha with some sweets as offerings. Wangu left a ring and a prayer flag; the Everest had blessed them and they thanked her profusely. From the Pumori Ridge, Bull could clearly see them reach the summit and passed the news to the other camps. But they only received confirmation from the pair by 6:00 pm when they dragged their limp bodies to Camp-VI.

Next day, the Sonams reached the Advanced Base Camp. While, Cheema and Gombu were welcomed at the Base Camp with decorated banners made from marking flags, they were received by a procession at the crampon point like heroes. Cheema danced and Bull along with others joined him. The whole camp was alive in celebration.

Vohra and Ang Kami were next in line. With the blessings from the weather gods, they left the camp at 6:00 am on the 24th of May. They climbed under a clear sky; their movements were closely followed from every observation point. By 8:50 am they were on the South Summit. Bull's eyes were trained on them and he followed their every step like a hawk. Walkie-talkies were kept at hand to catch any dramatic moment: and then, they did it.

"They are on the top," Bull reported to Kohli at the Advanced Base Camp, "Congratulations." Vohra had finally fulfilled his life's dream. He was in all the three expeditions and had earned his place on the summit. Later he told Bull that Ang Kami wanted a picture taken with his poster of the actress Sadhna on the Summit, Vohra refused and they had an argument over that on the summit. Kami never really forgave Vohra for that.

On their way back, Vohra slipped and lost his ice-axe at the rock chimney. Their progress was slow and they returned to Camp VI at 4:15 pm, both of them were complaining that their feet were getting cold. Meanwhile, the two Sonams had reached the Base Camp and their reception was no less grand than that of their predecessors.

Vohra and Kami spent a restless night in howling winds with cold feet. The winds hadn't calmed in the morning, they waited for a while but decided to head down to South Col anyway, somebody would come up for them from the Camp there. Somebody did. Mulk Raj brought them hot tea and they were returned to civilization. Their injuries were declared not serious later and they were both able to walk on the return march.

Rawat and Bahuguna, BP Singh and Ahluwalia, forming the fourth and fifth summit parties respectively, were about to start. Due to the short climbing period now coming to a close, after consultation, Kohli and Bull decided that the two parties shall push for the summit together. Assistant Sirdar Phu Dorji was added to the summit party. He had done excellently during the whole expedition and when they found out that they had extra oxygen bottles for one more climber he was the obvious choice. Thus, they asked him to make the South Col in two days and join the summit party by the 27th.

The fourth summit party had left Camp-IV on the 26th of May. A short distance up the Lhotse face, BP Singh complained of pains in the chest. He had to, therefore, withdraw and return to the Advanced Base Camp. The others went on to Camp-IV and the South Col the next day when Phu Dorji joined them.

The 28th of May dawned bright and clear. Ahluwalia, Rawat, Bahuguna and Dorji occupied Camp-VI that night. The next morning at 5:10 am, Rawat reported that they were leaving. That day was the twelfth anniversary of the first ascent of the Everest by Hillary and Norgay and obviously they were all very excited. However, a short way from the camp, Bahuguna felt weak and exhausted. He had been suffering an itch through the night, later diagnosed as an allergic reaction. He was proving to be a drag on the team and returned to the camp.

Rawat climbed alone for a while but he later roped up with Dorji and Ahluwalia. They passed the South Summit at 8:45 am and in their excitement reached Chimney at 10:00 am. They walked the last few feet arm in arm to reach the summit together at 10:15 am; Ahluwalia placed his wrist-watch and a photograph of Guru Nanak on the summit. Rawat placed an image of Goddess Durga, and Phu Dorji placed a silver locket containing the Dalai Lama's photograph.

At noon Bull saw them descending to Chimney and by 3:30 pm they had reached the Summit Camp. The clouds had blocked his view and that time he had not been able to see them reach the summit and due to a breakage in the line they could not get a wireless confirmation from them either. Bull indeed was sitting on nails.

It was only the next day, when the team reached the Advanced Base Camp and Kohli relayed the information, then Bull let go of the breath he had been holding so long. They were successful and his logistics were correct, planned and catered. Had BP Singh and Bahuguna not withdrawn, they would have 11 atop the Everest. Needless to say, that expedition had many 'firsts'. It was first time three people stepped on the summit together, first expedition to place 9 climbers on the summit, a record that was intact for 16 years. Also, it was the quickest ascent to the summit by Cheema and Gombu from the final camp. And of course, Gombu was the first man to climb the Everest twice!

With adrenaline flowing in full force, Bull successfully faced the litmus test with flexed muscles- his only regret perhaps being his inability to be on top of the Everest because of his medical condition. Bull's joys knew no bounds. In his heart he thanked 'Goddess Everest' or 'Chomolungma' in the Tibetan and the 'Sagarmatha' in the Nepali. She tested their perseverance five years ago and that time she rewarded them, and an abundant reward it had been indeed! On the 31st of May, they were all down at the Base Camp bidding farewell to the mighty peak under whose shadow they had spent such long and trying times. The return journey was filled with farewells and goodbyes, first at Thyangboche, then at Kathmandu where they boarded two IAF Dakotas and reached New Delhi on the 23rd of June. There was a large crowd awaiting them at Palam. The Prime Minister, the Cabinet Ministers, officials, family, friends, neighbours and the list went on. They were honoured, humbled and touched by the reception and the feeling of the Everest conquest was still sinking in them.

Their stupendous efforts were rewarded by the Government of India and the Leader of the Expedition was conferred Padma Bhusan while the Deputy Leader and 9 summiteers were awarded a Padma Shri each. These awards were announced before the team even reached

Delhi. When they returned to Delhi, the Sport Ministry also awarded them jointly with the highest sport medal – the Arjuna Award.

After the success of the expedition, many functions were organized all the state governments gave them state awards and receptions. Immediately after the climb of the Everest, Captain MS Kohli, Bahuguna and Sonam Nyamgal were sent to America to work with Perry Bishop – an American Everester and a CIA agent. The ambition was to put nuclear device on top of Nanda Devi that could record the movement of the Chinese troops in Tibet. Somehow, the device could not be put on the top of Nanda Devi and was kept in the western side of the peak about 500 feet short. But landslide had taken its toll and the nuclear device was still under the mountain despite many search parties. Captain Kohli has written a good book on the subject called "Spies in the Himalayas".

CHOMOLHARI - 1970

After the Everest expedition, Bull was posted as the Principal HMI, Darjeeling. He really enjoyed transmitting his lessons in mountaineering as an adventure sport to young boys and girls. It was indeed honour to have Tenzing Norgay as the Director of Field Training and Nawang Gombu – who was the first man to climb the Everest twice, as the Deputy Director for Training.

Most of the Sherpa instructors had immense experience. Some of them were on Mount Everest with Lord Hunt or had carried stores for Hillary and Tenzing. All these instructors had practical knowledge and knew little theory. Thus Bull made lesson plans for each class but notwithstanding, students could ask awkward questions to Sherpa instructors who had precarious theoretical knowledge base. He was sure that when educated Sherpas would come as instructors, these lesson plans would be very helpful to them.

The HMI was established by Pandit Jawaharlal Nehru and Dr. BC Roy – the then Chief Minister of the West Bengal. The expenditure of the Institute was shared both by the State and the Central governments. Most of the Principals came on deputation to that Institute; until then service officers had been appointed as the Principal of the HMI. Tenzing Norgay remained the Director of Field Training till he died at the age of 74. Today, the Institute is run by children of Sherpa instructors who are more educated as compared to their parents.

During his stay Kumar added an Everest Museum, apart from the existing mountaineering museum that depicted Himalayan climbing.

He also exhibited Everest films in the auditorium at a nominal fee. Once when Mrs. Gandhi was expected at the institute, the then the Governor of the West Bengal Padma G Naidu was very worried. In her honour, Bull had organized a huge rock climbing demonstration that was attended by thousands of people including Sunil Roy, the then Director General of Tourism who picked Bull from there to start the first National Ski School in Gulmarg.

One day while Bull was still in HMI, His Majesty Jigme Dorji Wangchuck, the third Druk Gyalpo (king) of Bhutan visited Darjeeling. He received him at the HMI gate with Tenzing Norgay. It was an honour to welcome the King who had opened Bhutan's doors to foreigners for tourism and growth. Bull showed him around the Everest Museum and the mountaineering films of their expedition to the Everest. His Majesty was kind enough to accept the invite to have lunch with them; it was while they ascended the steps to his Darjeeling home that the King said the magic words that left Kumar mesmerized and flabbergasted, "We have mountains in Bhutan, why don't you climb them?"

"I would love to," Bull replied and hesitantly added, "what about Chomolhari?" Now there was speechlessness on the other side of the table. All the ministers in the Royal entourage seemed to be thinking the same thing- "an Indian officer requests the King to be permitted to climb their precious mountain that is worshipped next to the Buddha in the kingdom. How dare he?" The tension in the room could be cut with a knife till His Majesty finally answered, "Only if it is an Indo-Bhutanese Expedition"and quickly Bull replied, "Of course, we would have it no other way. Thank you, Your Majesty" and requested the VIPs to move to the dining room for lunch.

The events that followed were quite a roller coaster ride. The lack of equipment was quickly resolved by the king's help and the 3 Bhutanese officers chosen for the mission were trained at the HMI. During the preparations of Chomolhari expedition he was the Principal of the HMI, so it was his responsibility to lead as well as prepare the team.

Outside the walls of the HMI, things were not so rosy. The letter from the Queen of Bhutan was just one of the many hurdles faced; but it was diplomatically replied to by the Indian Government stating, "We cannot interfere in the internal affairs of Bhutan". Tibetan Buddhists believe that Chomolhari is one of the goddesses, Tseringhma, bestowed with the charge of protecting land, the Buddhist faith and the local people by Guru Padmasambhava. Chomolhari is a holy site, where pilgrims pay homage on their way from Phari Dzong to the holy lake, Jomo Lharang, located north of Chomolhari. The lamas, driven by their devotion for the mountain goddess, were rousing superstitions about the dire consequences of the expedition. Their claims were fueled by the death of the Sherpa Sardar at a construction site. Thankfully the king and Bull's team spirit were undeterred.

Their group arrived in Thimpu on the 5th of April 1970 when the Paro Tshechu festival was underway. Though the festival was taking place in a fortress called Paro Rinpung Dzong near Paro city, one could see the festive signs everywhere. Celebrated during the onset of spring, Paro Tshechu is a 5 five day procession during which the monks perform religious dances. It was a pity they couldn't travel to Paro to experience it, and seek blessings. But perhaps, amidst the religious whirlpool, the expedition had whipped up some rumours and it was best they stayed away.

Thimpu was covered in colours and the festival did have some effect on the spirit of the people even from a distance. Besides, the city was preparing for the forthcoming visit from the Indian President. The energy in the air only strengthened their spirits and fueled their desires for success. His Majesty charged them with laying a "Yangu" offering on the summit of the mighty mountain. The pot, blessed by a Lama, contained gold and other offerings to the protector of Bhutan, Chomolhari. The expedition caravan of 10 comprised Bull, Dorji Lhatoo, Lieutenant Chhachu, Major Prem Chand, Arora, Captain Dharampal and Captain Kang along with Sherpa Nima, Gyalbo and Nawang. They left Thimpu on the 7th with a clear motive in their minds and hearts that the Indian Tricolor and the Bhutanese Dragon should be on the Chomolhari Summit by 10:30 am on the 23rd of April, when the Indian President would arrive in Thimpu.

They drove to Drukgyal Dzong, or what remained of the once mighty fortress. The crumbling walls of the ruins indicated an ancient history. It was built in the 16th century to commemorate the union of Bhutan and since then had been the venue for many incidents from diplomatic meetings to violent wars. It was a pity the structure was destroyed by the fire in the early 1950s. Even in its current state it seemed like the perfect watchman keeping an eye on those who dared to violate Chomolhari.

From there it was a long march to the Base Camp at an altitude of 15,000 ft. That had been one of the most peaceful marches in Kumar's career of mountaineering. Lack of problems from porters or muleteers gave them enough time to admire and relish the natural beauty that only the Himalayas could offer. Paro Chhu (river) flowed peacefully, white peaks loomed above and the occasional herd of *takin* grazing on the hillside shrubs was fascinating.

For the expedition they had planned to set up 3 camps along the path - the Base Camp, followed by Camp-I half way up the mountain and then Camp-II a few thousand feet short of the summit. The camps would be resting places for the mountaineers, making the climb a little more bearable and provideing shelters during nights.

On the 13th of April, as the caravan marched towards Jangothang, they crossed Tegethang, the winter home for Yak herdmen. Apparently then in Bhutan, a man's wealth was judged by the number of yaks he owned, an interesting fact to ponder while they looked for a place to set up the Base Camp. The heavy cloud cover and the absence of villages made it very difficult to pick the right spot. At around 2:30 pm Bull just stood still, his sixth sense and he halted the caravan to wait for the cloud to clear. Finally, when the new day dawned and the skies were clear, they found to their astonishment that the existing location was perfect for the Base Camp!

From the Base Camp to Camp-I, the serpentine path circling around the rocks and the glacier snout (or the right fang as Chapman called that) was too difficult for the horses; and there the yaks came to their rescue. The glacial tongue was smooth and unbroken, but the hurdle in their path was the labyrinth of ice. Lhatoo slipped into a crevasse and meandered his way through its ceiling gaping at the clear view to the top.

By the 20th of April, they had set up Camp-II at approximately 22,000 ft. Their goal of planting the flags by the 23rd of April seemed a little closer then. Motivated by the euphoria of winning the race, everyone worked hard towards setting up and stocking Camp-II sufficiently for the summit party to make its bid.

On the 22nd of April, Dorji Lhatoo, Chhachu, Major Prem Chand, Sherpa Nawang and Arora occupied Camp-II. After spending a comfortable night in two tents they left for the summit, crossed the yellow rock area and followed the South West Ridge. Just 50 ft below the peak, Chhachu was overwhelmed by the situation and his own religious sentiments towards the mountain and he could not go any further. While Chhachu thought the sacred goddess should not be trodden, his team mates believed they were merely pilgrims meeting the divine deity in prayer and bringing the "Yangu" as an offering.

Respecting his conviction, the others left him there properly anchored to an ice axe for his own safety. Bull witnessed all that from the camp, wistful that he could experience first hand that historic moment. But as leader he let his juniors grab the opportunity and conquer the untouched peak. He saw the party reach the summit. The time was 1000 hrs, the goal had been achieved, 30 minutes later the Indian President was greeted by His Majesty with the news that the Indian Tricolor and the Bhutanese Dragon flags were flying together in peace and friendship atop Chomolhari. Even the Prime Minister Indira Gandhi sent her compliments and regards via the Indian Embassy on the 26th of April 1970 that read, "Delighted to learn of the success of the Indo-Bhutan Expedition to Chomolhari Peak(.) Congratulations to the members of the team (.) Personal regards to Your Majesty (.) Indira Gandhi (.)"

The camp was alive with excitement and celebration, especially when the party members returned safe and sound. Encouraged by the initial success, the second party of Captain Dharampal and Captain Kang with Sherpa Nima and Gyalbo prepared to make the climb. That party left the Camp-II at 7:00 am; a rather late start, but made good progress. Gyalbo was not feeling too well and he stayed behind. Bull monitored their progress and by 9:00 am, they were on a ridge just short of 23,000 ft. He started making a film of Majors Prem Chand and Oberoi skiing to Base Camp.

When he turned his attention back to the mountain at 9:15 am, his heart skipped a beat, "Where are they?" He did not realize the words slipped from his tongue, Lhatoo snatched the binoculars from him and replied, "They are not there." Further speculation on the matter was halted due to a descending cloud that blocked their view. They gave each other hope; perhaps they were behind a ridge hidden from their view and when the clouds would clear they would be able to see them. The clouds did not lift and the entire evening Camp-II remained obscured to eager eyes waiting below. The night came, giving rise to more worries and doubts about their survival. That was the longest night that Bull had ever spent, the longest that he could remember. Tossing and turning, checking for the sun to rise every 10 minutes, cursing the lack of communication with Camp-II that was a relentless restless night!

When sun finally peaked over the horizon, Bull rushed out of his tent and looked up. It all seemed peaceful, silent. He noticed a Sherpa coming out of Camp-II and after staring at the summit for a long time started waving frantically. The tragedy was confirmed. Captain D Pal married only four months ago and Captain Kang father to a four months old baby, were perhaps no more. Nor was the Sherpa whose only relation in the world was his old ailing mother. Lhatoo with another Sherpa reached Camp-II and took Gyalbo down to Base Camp.

An aerial search was carried out by the Indian and Bhutanese Forces that proved futile. Lhatoo, the other team members and the Sherpas went down to the Base Camp but Bull stayed at Camp-I as he did not want to leave till a proper search was carried out.

Another piece of drama played itself out at the Base Camp. A Lama had paid a visit and announced that the tragedy was a result of their foolishness to ignore the predictions. He went on to conclude that goddess Chomolhari would not be calmed so easily and anyone who went to search for the bodies would meet with a similar fate. That seemed to shatter the morale of the Sherpas and crew members. They requested Bull to come down and reconsider his decision after counselling with the Lama.

If he returned, then coming back would not be an option. The luxury of Base Camp would be too tempting to allow anyone to leave and return to look for his men. No, he would stay there until they found them. If the Indian Air Force (IAF) was not successful then he would go and scout the area himself and volunteers could join him to climb to Camp-II.

Bull was glad when Major Prem Chand and Sherpa Nima arrived at Camp-I. Together they moved up to Camp-II; and the next day ascended to the ridge where the accident must have happened. Chomolhari stands on the Tibet-Bhutan border and they had climbed from Bhutan.

Prem gave the following account of that search in his diary - "*With one Sherpa carrying the oxygen cylinder, we were on our way to the peak. Snow enroute was harder than it was on the 23rd. We kept on advancing, breaking the trail till we reached the South West Ridge. From the top of the yellow rock we could see Camp-I below. After some rest we went on. I was leading the rope, followed by Sherpa Nima and the leader brought up the rear. From the yellow rock we kept the ridge 4-10 yards to our left. Our progress was slow due to the altitude and the fatigue we were experiencing. As we approached the second yellow rock area below the south summit, I halted to catch my breath. My position gave me a beautiful view of Phari Dzong. The Sherpa was bent over his ice axe, catching his breath as well. The time was 8:20am.While I was catching my breath, I heard a sound. As if a fast moving object had pierced the snow over the ridge. I saw a small piece of ice uprooted from the mountain side and some powdery snow was blown towards us. I was shocked and was still examining the piece of ice when the realization hit me. "A bullet!" I wondered out loud. I called out to the leader, "Sir, a bullet has been fired at us!" He was examining the area at a lower region, looking for some clue of our mates. He asked me to hold on to the piece while he came up. I showed him the ice and explained the sounds. Even the Sherpa had heard the sound and seen the piece fall. "But at that height?" the leader was still doubtful. "It seems to have come from the Tibetan side, Sir." I explained. We wanted to carry on our search to the peak but looking at the circumstances, we decided it was best to return.*"

Prem Chand and Bull returned with disturbing thoughts clouding their minds. Could it be possible that the Chinese shot the party down? Were they taken prisoners? They did not wait for an answer and made their way back; at least then Bull could face the families of the dead and tell them that a sincere and thorough effort had been made to look for the climbers. Though the Chinese Government denied any suggestion of action on their part against the mountaineers, Bull's heart sinking to abysmal depths of despair was not convinced.

Well, in the end the Lamas proved true. They faced the wrath of the mighty Chomolhari, and lost great friends and officers to her anger. They said it was the presence of the "Yangu" that saved the first party from any mishap, but since the second one was not protected by the Lama blessings and suffered wrath of the goddess.

A few days after the expedition, an Indian Airlines aircraft crashed in the Bhutanese territory. The Lamas called it the aftermath of the Chomolhari expedition. Again a week later when the President of India was returning from his visit to Bhutan, a falling stone killed one of the correspondents. That too was attributed to the scar on Chomolhari's chastity. Ignoring the superstitious claims, His Majesty Jigme Dorji Wangchuck, was very sympathetic to the misfortune met by the expedition. But he did say that the names of the deceased as well as of those who reached the summit would go down in the Bhutanese history along with the details of their expedition. While Chapman might had succeeded against odds in his time with his tweed jackets, canvas tents, nailed boots; the hemp rope and bamboo ladders were quite useful. The success and the celebrations would have been grand if not for the losses of the brave officers and mountaineers in that expedition.

The Chomolhari was conquered for the first and the last time in 1970. No more expeditions were permitted along the Bhutanese side of the mountain since then. Though trekking is popular on the mountain among tourists, they are limited up to the Base Camp altitude. Above that, Chomolhari has remained untouched for decades. Though the Bhutanese Tourism Department permits treks in the Himalayan range within its borders, no one is permitted to climb Chomolhari beyond the Base Camp altitude of 13,251 ft ie. Jangothang. Fortunately for them, the tourists get a grand view of the mighty mountain from that location and the agencies make a decent living. But it certainly is a pity that such a beautiful mountain remains unexplored. Bull was thankful for the opportunity he had and paid his homage to that ever powerful goddess.

Crossing Sun Kosi River in the dug in canoe

Advance Base Camp - Everest

Members of the 1960 Everest Expedition. Leader Brig. Gyan Singh in the center

Reception of the team of the 1960 Everest Expedition at Palam Airport

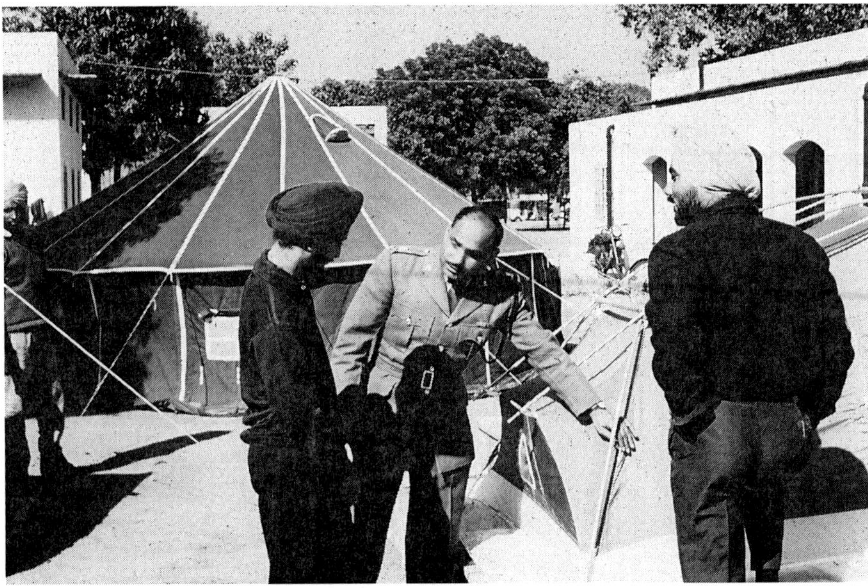

Maj Kumar and other member of the 1965 Everest Expedition inspect the equipment

Members of the 1965 Everest Expedition with Shri Lal Bahadur Shastri, the Prime Minister of India

Reception of the members of the 1965 Everest Expedition at Palam Airport. The whole cabinet along with the then Prime Minister Gurzarilal Nanda

Col. Kumar with Junko Tabei, the first lady who climbed the Mount Everest

With Jigme Dorjee Wangchuk, His Majesty - the King of Bhutan discussing about Chomolhari expedition

On the summit of Chomolhari

Col Kumar briefing Gen. T N Raina, MVC regarding Kanchenjunga expedition

The expedition did not step on the summit of Kanchenjunga which is worshipped by the Sikkimese. They remain 6 feet below and planted the Tri Colour

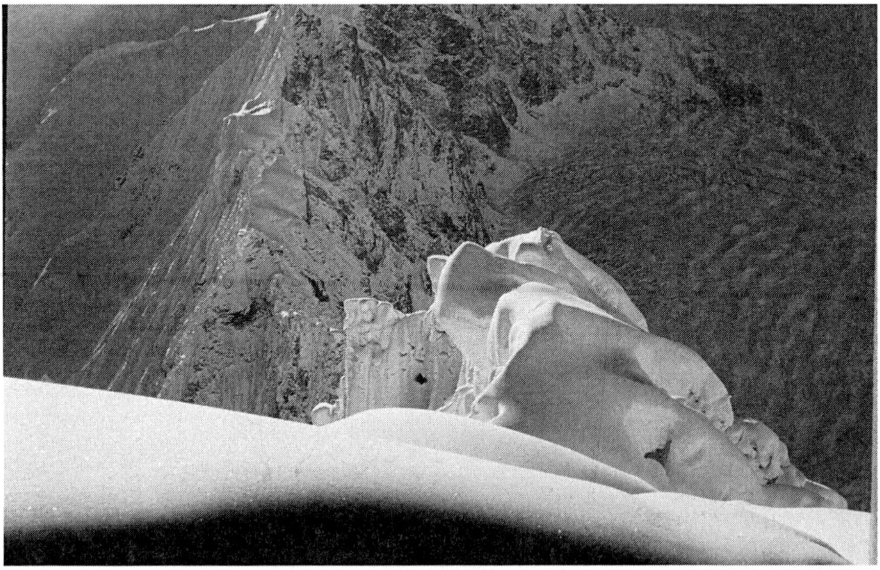

The North East Spur – the main obstacle of the expedition

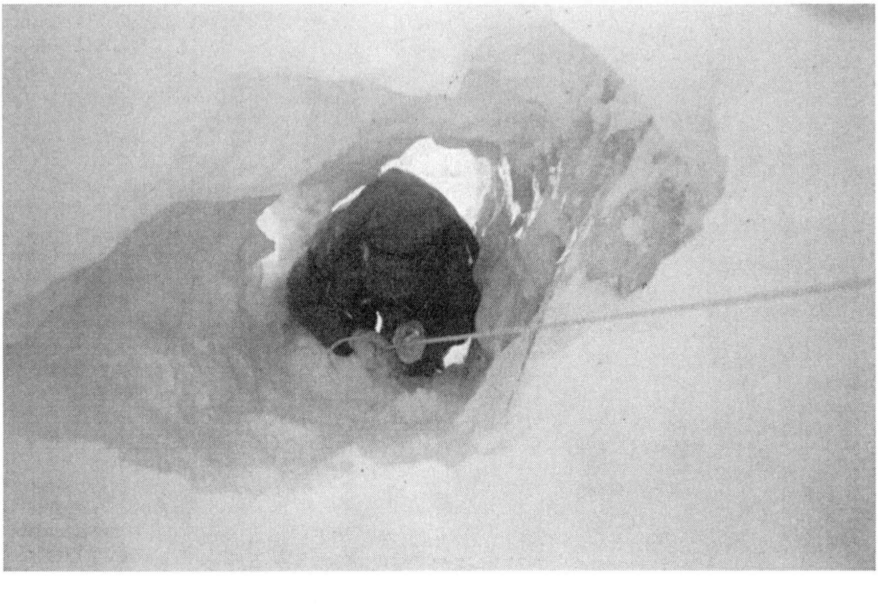

Sometimes the team had to dig tunnels through the spur to find a better route towards the summit

Col Kumar and Prem Chand at Camp V

In 1983, Kumar led a successful expedition to Kamet – the 3rd highest peak in India

NATIONAL SKI SCHOOL

After the Chomolhari expedition, Bull was shifted to Gulmarg to train the first batch of Ski teachers at the National Ski School now known as the Indian Institute of Skiing and Mountaineering. He was the Principal and added mountaineering to the syllabus so that the boys could be employed as ski teachers as well as mountain guides.

It was a very interesting position to hold, but to start an Institution from scratch there were endless teething problems. They did not have any transport and infrastructure. The syllabus had to be made for 3 years and all financial demands that went to the Tourism Department were in turn sent to the Finance Ministry for approval; and it took ages to get any sanction. He remembers that they had a small chair lift for taking skiers up the slope that had developed some technical snag and they need Rs 5,000 to put that right. Though he was the Principal he had no financial power. He knew that the relevant sanctions would come only after winter. So, he spent his half month's salary so that they would not lose the entire winter season. The sanction did come after a year and he recovered his money.

During his tenure in the National Ski School, Bull got the United Nations Fellowship to learn ski teaching techniques in Austria and Switzerland. At that time a new technique in skiing had been innovated called-"Jet Austria". That technique was very useful in tackling the moguls that get formed up on the ski slopes. It had been masterminded by Prof. Krauker Hauser – Head of the Austrian Ski School.

When Bull was in Austria, he met a Punjabi from Pakistan. Though India and Pakistan had been at war in 1971, both as Punjabis had a great time together. Bull made very detailed notes on his training so that he could introduce new methods in the National Ski School. But as no infrastructure was available, he could not incorporate those instructions at the school.

Once, when he was in Switzerland at Davos Ski Resort, he saw four alpinists who were preparing for a weekend jaunt on the slopes. As thousands of people ski, he thought these alpinists might know the Himalayas. He started a conversation with them and asked them if they had been to the Himalayas. One of them replied in the affirmative. Then he asked him about the Indian Himalayas and he was surprised when he said that he had also met Colonel Kumar – the Principal of HMI, Darjeeling.

So, when he told him that he himself was Colonel Kumar, the alpinist cancelled his trip for the weekend; and he and his wife took leave and looked after him as their guest He reminded Bull of the incident in Darjeeling when His Majesty the King of Bhutan was having lunch with Bull he somehow had managed to come to his residence and sent a small paper saying that he was an alpinist from Switzerland and he wished to meet him. Bull went out and found a man in torn jeans with mountaineering equipment. For the King's lunch, as per the protocol instructions of the Indian Government, they all were ceremonially dressed in 3 piece suits. Bull invited him to join them for lunch, but he protested saying that he was not dressed appropriately. Bull told him that it was a mountaineer's house and he was dressed for that; and took him inside.

It so happened that the King of Bhutan – Jigme Dorjee Wangchuk used to go to Switzerland for hot treatment for body pains in the same town in which that Swiss alpinist lived. They had so much to talk about that Bull had great difficulty in introducing the other V.I.P. guests to His Majesty. So, he was very pleasantly surprised that a small gesture on his part by giving him lunch was returned with one week of hospitality!

While at the National Ski School, he also introduced water skiing in the Nagin Lake. The boys were kept in house boats; it was a great life indeed. Information Ministers of the non-aligned countries were coming to Kashmir for the first time and his School was asked to give some demonstrations. Apart from showing various techniques, Bull also had a shikara race and when the shikaras came to the finish line, they picked up the shikara that had come first. But there was a huge commotion as there were about 50 shikaras shouting at each other. Mohammad Abdullah, the then the Chief Minister of Jammu & Kashmir (J&K) instead of being embarassed told the other foreign dignitaries and Information Ministers, "Look at the people I have to deal with!"

INDUS BOAT EXPEDITION - 1975

A few months later, Dr. Poncar and his friend Dr. Volker Stallbohm had visited the Chief Minister of Jammu & Kashmir Mohammd Abdullah with an idea of rafting down the Indus River. Jaroslav Poncar, a 30-year-old West German had hit upon the novel idea of paddling down the River Indus while he was traveling by bus from Leh to Srinagar during his previous visit to Ladakh in 1974.

He and his friend Dr. Volker Stallbohm had earlier paddled rivers in Mexico and Guatemala and the thought of doing a virgin river, one of the five big rivers in the world -- something unique, haunted them till they landed at Leh one fine morning. The Indus expedition came to Bull's way by chance. Had it not been for Sheikh Abdullah's keen interest in him and his activities, he would not have had that magnificent opportunity. The Chief Minister sent those two Germans to Bull and they told him about their plan to raft down the Indus. He was very enthusiastic and preparation for the adventure commenced right away.

The 2nd of August 1975 was to be the D-Day when they would start their trip down the River Indus. They left Leh on the morning of the 1st of August and reached Chumathang in the afternoon. Kumar spent most of that journey thinking about the various problems and hazards that they might face on the Indus.

Of all the hazards that he could think of, cold was probably the most serious. At freezing temperatures, they could be numb in just 3 minutes and become ineffective in another 5! He tried thinking of all the possible ways he could protect his frostbitten feet, for ever since the amputation of his toes, they had become quite susceptible to cold. He normally used a pair of battery heated socks operated by ordinary torch cells. But he knew that once drenched in the icy cold water of Indus, both would be quite useless. Obviously, he would have to think of some other method to keep himself, and particularly his feet, warm.

Then there was the hazard of the boat getting punctured in the river and capsizing. He thought, however, that not only were there four separate compartments, but the Germans would also be carrying a puncture repair kit. They seemed to know their job and since they had selected the boat, they must already have worked out ways to get around such a problem.

On the river they would have to battle against whirlpools, hollow rocks, rapids and waterfalls. A whirlpool, he thought, would have to be really strong to suck in an inflated boat such as theirs. As for the other three hazards, those were risks that had to be taken, and any such expedition as theirs always depended on luck to some extent. If they were lucky, then they would avoid such problems; if they were not, well then, too bad. Anyway, it would be best to tackle any such situations as they arose, for it was not really possible to anticipate most of them in advance, and be forewarned and thus forearmed!

One incident from the Teesta expedition that he had led earlier, however, kept haunting him. They had got stuck among boulders, in midstream, and had found the gushing channels on either side quite unswimmable. On the Teesta, such a situation was not so bad, since they were always in sight of the road and a rescue attempt from there was possible. On the Indus, however, they would at times be doing a 40 km stretch where the road was far away from the river, not even visible from midstream. If anything untoward were to occur on any such stretch, then surely all would be lost. But he did not let such thoughts worry him as worrying was quite futile.

Before starting from Leh, Bull had checked his equipment. He had carefully gone over his battery heated socks, hunter shoes and his crash helmet. The helmet had been presented to him by Fritz Morevec of Austria, during the International Mountaineers' Meet held around 2 years ago. He was the only team member with a crash helmet. He also checked the two life jackets he would be taking along. One of those had been borrowed by him from Snow and Jungle Survivor School of the IAF and was the kind used by the pilots while bailing into the sea. Light and easy to inflate, its special feature was that it would at all times keep the head out of the water. One disadvantage would be that such a jacket could be easily punctured, and in that case he would certainly be doomed.

In Leh, just before he had arrived there, the others had gone on a trial run wearing shorts. That had given them bad sunburns. The body would thus have to be kept well covered. Warm clothes for that purpose would be quite useless. Not only do they become heavy when wet, they also take a long time to dry. That would be a disadvantage for it had been proved that the body loses 240 times more heat if the clothes were wet. Also, to protect against dryness caused by the chilly winds and cold waters, he was carrying a bottle of Vaseline.

They started for Chumathang only after they had finished checking their equipment, after they had breakfast with the Development Commissioner Mehmood at his lovely house that was decorated in Ladakhi style. While eating that lavish breakfast, Bull had felt quite the proverbial *"Id ka bakra"* being properly fattened before the kill. He was quite apprehensive about their trip and was wondering whether they would ever come back alive from that treacherous expedition they were about to embark upon!

After reaching Chumathang, they pitched their tents first and then they looked at the river. Jarav expressed the desire to start the trip from 40 km upstream after seeing the river. Since they had not taken permission for that, and that would be against the Government regulations, Bull somehow managed to dissuade him from trying to do that. In the end he agreed to start from Chumathang as had been planned. Chumathang at 13,800 feet was quite cold and uncomfortable. He allowed the team members only two medicinal quotas of brandy each. Later, he went and bathed in the famous hot springs of Chumathang that really relaxed and soothed his edgy nerves. On his way back to the camp as it was getting dark, he had one last look at the river Indus before settling down for the night.

Bull woke up early on the morning of the 2nd of August 1975. He was excited and anxious to get going. The morning was cold and cloudy, and the prospect of the icy cold waters of the Indus was not pleasing, but the team was not likely to put off its start. As usual, he lazed around in bed having his cup of tea while the other team members hurried about their chores. For, despite the fact that he had led an unusually energetic adventurous life, he still enjoyed spending nearly an hour in bed every morning, lazing and brooding. For him that was possibly the most precious part of the day. It gave him the wonderful feeling of being one-up on everyone. When he stepped out of his tent it was to find the others almost ready to leave. Hastily he went back into the tent and dressed up speedily.

When he re-emerged, it was to see the Lamas who had assembled on their own to pray for their safety. It was a touching scene and he was moved to see them chanting their prayers even though he was hardly in a mood for any religious ceremony right then. He still had to pack his rucksack, load a film in his camera and have breakfast. Quickly, camera clutched in one hand, jacket and helmet in the other, he rushed after the others for the launching site. Jagmohan followed him around carrying a plate of breakfast for him. Before Bull could refuse, he said, "Doctor's orders!" Since Bull could not possibly refuse, he wolfed down a quick breakfast. He was reminded of General Carriapa's advice that before starting a day, breakfast was non-negotiable.

He reached the launching site only to discover that the site had already been shifted a bit by the others so that they could have better light for photography. Both Jaroslav and he decided, however, not to take out cameras on the boat. He gave his camera to Jagmohan, having first set it for him to take pictures since he was quite a novice at photography.

Finally, Kumar decided that the time had come for them to start. He felt elated, and his spirit soared at the thought of riding the waves of the mighty Indus. But he was afraid as well; and very conscious of emptiness in the pit of his stomach. While some people claimed they never felt pangs of fear, he was quite sure that that was not true. What they probably meant was that they had overpowered fear; or else, never showed fear if they felt it at all. Fear was a basic human emotion experienced by all living beings at some point in time. However, the brave amongst us are the ones who can master it and control it, not allowing it to deter them from the objective that needs to be achieved. It was a strange feeling though, that simultaneous feeling of joy and fear, while he walked towards their boat – the Helena Dolma, lying by the side of the river.

They slowly and gingerly lowered their boat into the river. Softly it touched the surface of the water, and lay there calm and steady. It was a black boat. The Indian and German flags on it gave it a dash of colour. The boat looked handsome, inviting; anchored there as it was to the bank, with a long thin orange nylon rope.

Pali stepped in first and took his place at the front left cover. He fiddled around with the flags, re-arranging those as he waited for the others to join him in the boat. Bull went in next, almost tripping four water bottles and a foot pumps to be used for inflating the boat and the life jackets. He noticed that these had not yet been tied on the floor and he ignored it thinking that Jarav would do the job.

Stall Bohm or Volker as they called him, stepped in and took his place behind the steering. That, Bull thought, was odd, since in the trial run it had been Jarav who had been handling the steering. Then Jarav jumped in and shouting "let's go" put his hand on a rock and pushed away from the bank. Amidst shouts of "best of luck" and "*Indo-German expedition ki jai*" from the crowds that had come to see them off, they commenced their onward Indus journey.

Soon they were in midstream. The powerful water of the Indus carried them along, ahead lay the 200 km stretch that they planned to cover over the next few days. Bull turned around and looked at their crew. He noticed that four religions were represented in that expedition. There were Jarav and Volker who were German Jews, Pali - who was a Sikh, Aziz – a Kashmiri Muslim, and of course himself – a Hindu.

They had barely moved a hundred meters when suddenly the boat sank under him and turned over. Kumar was thrown into the icy water. He tried to surface but could not. Desperately, he tried once again and failed. He realized that he was trapped under the upturned boat. Fear gripped him then. The freezing waters, flowing hard and fast against him were already numbing him and he could not possibly hold his breath longer. He had to do something emergently. Using all the might that he could muster, he pushed the boat away and put his head above the water. Gasping for breath, he clung on the boat. He found the others too doing the same.

He saw something orange floating away on the waves, some distance away from the boat. Numbly it registered in his mind that it was Pali's life jacket. He had not yet worn that when the boat overturned. Fortunately, he noticed that Pali was hanging on the boat. Otherwise he would have been in serious trouble.

He thought that if they were to be stranded in this manner for even another five minutes they would be numb; rendering their arms and legs ineffective, thus enhancing the possibility of drowning. He urged everyone to try harder. They did, and miraculously made some headway against the magnetic pull of the powerful currents. All of them struggled hard, and such activity probably kept their blood circulation going, saving them from death by freezing! Slowly they moved towards the bank.

After what seemed an eternity, and just as Bull was losing hope once again, Pali called out that he had struck a rock with his feet. Digging in quickly, he helped the others find their feet once again on solid land. They had survived their first crisis.

The crowds moved towards them and offered whatever help they could. Thoughtfully, they gave them warm blankets that they were themselves using. However, before wrapping up in the warm folds of the blankets, they turned over their boat. It was then that they discovered that the water bottles and the foot pump were missing! Untied, these had obviously fallen in to the river. The loss of the pump dismayed Bull. He was happy, however, that everyone had survived. He was also glad that he had not carried his camera, for that too would have sunk during the accident.

They stood shivering on the bank of the river. Pali had been hurt. He was bleeding, but felt no pain, numb as he was by the freezing waters. Jagmohan – the expedition doctor attended him. Bull gave Pali his spare life jacket since he had lost his own.

They discussed the cause of the accident. Jarav Poncar had given Volker Stalbhom the steering, trying to save his own energies for the rapids that were to come later. Volker, by no means an expert helmsman had been unable to control the boat while others were shifting around and settling down. So they decided that, for the rest of the trip, Poncar would steer the boat.

The start had been bad. They decided, however, to carry on right away. Though less confident that time, Bull was sure they would be more careful and would try harder.

Once again they boarded the boat. Pali took out a pocket sized Granth Sahib and read a few verses in prayer. Then he touched the holy book to his forehead. Religious rites over, he pocketed the book. A hard push and they were off once again.

For about 10 km the going was smooth, just as Jarav had anticipated. At other places where the water was deeper and quieter the going was even easier. Looking ahead Bull spotted in the distance the crowd that had gathered on the banks. At the same time he also became aware of a change in the sound of the waters. It had a higher and deeper tone to it. Then suddenly he realized what that meant; and why the crowds were standing there. Obviously, there were rapids ahead.

Sure enough, as they took a slight turn around the curve of the river, the rapids came into sight. Stretching over 500 meters or more, these were dangerous rapids. The foaming, brawling waters seemed to jump 6 feet and higher. The question was, could they make it through these rapids? They would find out soon enough. They entered the rapids paddling hard, trying to steer out the boat towards the right bank where the waters were relatively calmer. But even these were quite bad! Their boat was tossed about mercilessly. They struggled hard to keep its balance. They seemed to be doing alright when exactly what Bull had feared happened. The boat ran into a rock and the force of the water jammed it there unable to budge.

Helplessly he looked around for some way out of the situation. It was then that he saw the rock as the frothing water receded off for a moment. It was just a few feet away. He leaned out of the boat and placing his hand on that rock pushed hard. The boat tilted a bit but settled down again. Pali, seeing what Bull had been doing leaned out to lend him a hand. With his longer reach, he was able to apply a good amount of force. With their combined efforts, they managed to push the boat away. Adventure sports are much about anticipation, group cohesiveness, skills and physical as well as mental fitness.

As they came off that rock, the force of the water pushed the boat around. The worst of the rapids lay ahead. Entering backwards would have been suicidal. "Try turning it around!" Jarav Poncar shouted. They had just about managed that when they were carried into the hellish fury of screaming waters!

With stunning force they were tossed around in those waters. The turbulent waves would smash into them at high velocity. The boat began flooding. Its bottom hit a rock and one of the wooden strips of its floor snapped, hitting Bull hard on his foot. Half blinded by the flying waters, desperately trying to drain the water from the boat, being jostled about harshly, and trying to paddle and steer the boat all the while, the crew seemed in the grips of a nightmare.

Jarav Poncar shouted, 'We bank on the right!' furiously then they began paddling as he tried to steer the boat towards the bank. 'Harder! Harder! Harder!' Jarav urged them on his voice cutting like a whiplash; and they reacted by paddling with all their might. It was obvious that they could not last very much longer. Wet and cold, their energies depleted, they were tiring quickly, very quickly but they made it to the bank.

The water was intensely cold. Bull looked down at it. It was sparkling clear and vibrant. He could see the sharp stone on the floor. Those hurt him every time he stepped on them. But duty had to be done. The boat had to be emptied. Somehow he suffered through it all until the boat was emptied of water.

Pali was still suffering the after effects of the jarring experience. But he recovered quickly and without consulting our doctor, he declared himself fit to carry on. While on the boat Bull had been unable to use his battery heated socks and jacket but he put these on then. Warmth flowed into his body and immediately he began to feel better. He walked back towards the rivers to look at the rapids that they had come through. From the bank, these powerful roaring rapids looked dangerous indeed. His confidence in their boat – Helena Dolma that had so successfully stood up to these rapids, increased. He then took a number of photographs of the rapids.

They spent hours resting and then decided to have another go. They carried the boat away from the rapids and into deeper waters. The portion of the river they were then in, was calmer and steadier. The river had spread out and slowed down. They barely felt the current under them. Bull was now beginning to understand the river. He knew that this was merely the calm before the storm. They decided to bank and reconnoitred before carrying on. It was a wise decision. A short distance ahead, the water went over a series of small falls before going over a sharp fall. Had they come to these falls unawares, they would certainly been in serious trouble.

"The boat will not take that fall with too much load. I would like to do it alone", said Jarav Poncar. Bull knew he was right, but he was reluctant to let him go alone. Bull asked him if it would be a good idea for at least one of them to accompany him. Jarav Poncar refused. As he knew his boat the best, and was also the most experienced and expert boatman Bull agreed to let him try it alone. Jarav walked back to the boat thoughtfully, shoulders stooped. Climbing on, he pushed away from the bank. Quite scared for him, they all watched as he approached the fall. Would he make that 10 feet fall successfully, they all wondered apprehensively.

He did, but not without giving them and himself some anxious moments. Keeping to the right bank of the river, he hit a rock as he approached the fall, and the boat turned around. It could have been fatal had he gone over that way, but fortunately he hit another rock and the boat straightened once again. The front portion of the boat went riding over the waves at the edge, and the boat landed right side up on the water below. He had done it.

The boulder that Jarav had hit, however, had done its damage. There was a 10 cm slit in one of the compartments of the boat. They had with them the puncture repairing kit but no pump since that was lost in their first mishap in the river earlier that morning. The valve in the compartment needed a special socket on the pump and an ordinary pump could not be fitted on. So they improvised a socket from whatever materials they had. It took them two hours to fix that puncture and yet that leaked. The reason, Volker pointed out, was that after a puncture had been fixed, the boat had to be left deflated for at least one night, otherwise it would not mend. So, the boat certainly was not in the condition to take any more beating that day.

Reluctantly then, the decision was made. They decided to earlier plan, that of boating from Pushi to Khalsi, a distance 150 km. On their first day then, they had only covered 17 km and were left with a damaged boat and physically, at least, a beaten crew! The river had not beaten their resolve, however, and they were ready to take it on the next day.

The 3rd of August 1975 was a bright and sunny morning. They prepared themselves for the hard day ahead and set off for the launching site below the Pushi Bridge. Bull stood in the water helping the Helena Dolma into the Indus. For the first time then he became aware, really aware of the waters flowing past. The moment was profound; the motion built into the waters by the composition of the land through which the river had passed over thousands of years. So vital, and yet so uncaring flowed the waters; and he was filled with wonder.

When they sat out that time there were only 3 of them in the boat – Pali, Jarav and Bull. They paddled towards Karu. It was pleasant smooth going. By now Bull had begun to understand the river quite well. He was able to make out the underlying hidden rocks by the subtle changes in the look of the water as it passed over them. At Karu lots of families gathered to greet them.

The next 20 km or so were a dream. The waters were calm, the wind behind them and they drifted peacefully along. The river was fast but there were no rapids. They passed the Tikse and banked at the Tse Gompa for a while as they were expected at Chuglamsar only at 4 pm. The Head Lama of the Tse Gompa had come down to see and bless them. They started for Chuglamsar at 3:30 pm and reached exactly on schedule. From a distance they could see that huge crowds had gathered to welcome them. On that second day, they had done quite well as compared to their first day on the river. They had stuck to their schedule and had covered 45 km as planned. Their boat however, was still leaking. So that time they got the puncture fixed the Indian way. They were assured that it would never give them trouble again, and right enough, it never did.

Next day, the 4th of August, they awoke to the chatter of people who had gathered to see them off. For 10 km the going was smooth and it seemed that everything would turn out fine; but then they met their Waterloo. At that point the river took a sharp turn to the left. They banked there to empty their boat of water before entering the fray of rushing waters curving around the twists and turns again. From the right bank then, they relaunched their boat.

Before they could enter the main channel, they were sucked towards a huge whirlpool at the mouth of a hollow rock. Along with the waves, their boat was pulled into the whirling inferno of waters. Emerging from the wave Bull found that the boat was empty except for him. Stunned, he felt boat toppling over to the left. Even as he threw his weight on the right, he understood the reason why the boat was tilting- Pali's legs were still entangled in it, while he was being dragged on the waters on the left side. Bull gripped his legs hard and tried pulling him up. He could also see Jarav's cap floating away but he himself reappeared on the side of the boat a moment later and hauled himself up into it. Together they pulled up Pali. Volker however was still missing.

Then they sighted Volker and terrified, they froze. The waves had captured him. Violently, brutally tossed about, he was being carried towards that hollow rock. The danger was that he might be smashed against that rock or get stuck into its hollow. Helplessly, impotently, they watched him struggle. Obviously, he understood the danger, for at the last moment he placed his hand on the head and prevented it from smashing into the rock. And then miraculously, he was flung away furiously by the waters, just as our boat had been. He landed near the boat and Jarav held him. We lifted him back into the boat. He was safe and the tension suddenly drained out. Bull looked up in relief, and once again froze.

There was a massive ten feet wave coming straight at them. Bull yelled, and they all looked. Before they could react, the wave was upon them engulfing them like a monster. It was almost as if the boat had taken off, for it practically flew to the crest of the wave and leaped off, falling what seemed an immense distance as the wave came down upon it. Why it did not topple was still a mystery to Bull. Anyway it did not. They found, however, that the boat was completely flooded. "Bank, bank" shouted Volker. Paddling madly, they banked on some rounded rocks.

They rested for a while on those sun baked rocks, and then they started again. Some distance later, they entered another series of rapids. There they were pushed uncontrollably to the left bank. Then something they had feared all along happened. The boat was once again punctured. There was a long slit along one side. They had to bank again. They did not know just how and when the puncture had happened. Jarav thought such a slit would only have been caused by a sharp steel rod. Bull thought that too much banking had something to do with it. Probably, some sharp, jagged rock, of the type that lay along the banks had been responsible. Anyway, it did not matter how it happened, what did matter was that they were stuck.

Earlier they had tried to plug the puncture with rope. But they had failed. But it was then that an idea had struck Bull. If they could stuff the punctured hole with dry cotton and plug that properly with cloth, on getting wet the cotton would expand and possibly become leak proof. The idea was worth trying and they did try that.

Tearing the pocket of Jarav's shirt made of hand woven cotton that seemed just right for the purpose, Bull had inserted that inside the inch and half long punctured slit. Then, still holding the ends of that piece of cloth on the outside, he stuffed in as much cotton into that slit as was possible. Tying together the ends and stuffing in more cotton along the edges of that slit, Volkar and he then lowered the boat into the water; and wonder of wonders, his idea worked! The puncture had been successfully plugged and the boat stopped leaking.

They started paddling again and came to the confluence of the Indus and Zanskar rivers. He was surprised to see that the Zanskar had at least as much, if not more, water as the Indus. They had to be very careful while entering that confluence, since there was a great deal of backwaters. Also, they had to enter the combined flow of the two rivers at a correct angle, for otherwise the boat would topple.

It was 6:30 pm when they landed near Nimmu where the road party was anxiously waiting for them. They had been lost to them for almost 7 hours. They told them that they had been very worried, particularly about the possibility that they had met with an accident along the way.

They all slept soundly in the night. The 5th of August was to be their last day on the river, since they decided to do the remaining 60 km to Khalsi on that day. Unfortunately, they were unable to do that. They banked at Alchi for a while and launched their boat again. At Saspul almost every villager had turned out to greet them. They were able to bank a km ahead of the point where they had actually planned to bank. Even though they had only done 32 km instead of the 60 km, they decided to call it a day.

Early on the 6th of August, the expedition party went down the river for a thorough reconnaissance of its course; lucky for them that they did as they discovered that the Nurla rapids had many sharp rocks jutting out of the water. So, it was decided that they would pull out of the waters before Nurla and re-enter only after the danger zone. Of the 17 km remaining after they had cut out the Nurla rapids from their schedule, Aziz, Volker and Jarav did 7 km. After that Pali and Bull replaced Volker and Aziz on the boat.

One km before the finishing point, they tried getting closer to the right bank so that they would not be carried past by the mainstream.

They banked in Khalsi at 12.30 pm The trip was over. They had boated down the Indus just as they had set out to do. And for the entire 150 km it had been a great going. Tough, yes, it was but laced with fantastic team work and immense fun. They had all really enjoyed themselves and had loved every single moment in the Helena Dolma.

Though the Indus Boat expedition was not a very big adventure, it got tremendous publicity as Emergency had been enforced in the country and the media could not publish anything unless it got censorship clearance. So, they were in the headlines for the next 4-5 days. In fact, when Mrs. Indira Gandhi came to Srinagar during that time, she called Bull to show her the slides of the Indus boat expedition.

There were 3 seats behind the projector and Bull was projecting the slides with Mrs. Gandhi on his right and Mridula, his wife on his left. Sanjay Gandhi came in and Bull told him that there was no screen. He looked around and saw a white dining table cover; he pulled it up breaking two pieces of crockery and hung it on the curtain rod making a perfect screen for Bull's projection. Halfway through, Maneka Gandhi – wife of Sanjay Gandhi walked in. As there was no seat available, there was havoc in the room as people were rushing to get a chair for her.

Bull till today feels little uneasy over the exaggerated and undue publicity they got for a small adventurous event and one publisher published his book on "Indo – German Indus Boat Expedition". He had selected two students of his for that adventure; but, unfortunately, the boat was too small to accommodate all and one of them was left out. He contacted the Blitz newspaper published by Karanjia and planted a story. His story was printed in the newspaper as "Spying down the Indus". That created unwanted chaos involving the Indian Army.

They had started their journey from Chumathang where foreigners were not allowed. He was hauled up by the Army for having Germans in the expedition, even though the Army's clearance had been accorded and the local Superintendent of Police had issued permit to enter the restricted area. Fortunately he had the permit with him issued by the Superintendent of Police on the instructions of the Chief Minister Mohammad Abdullah. However, they were put under

Intelligence Bureau (IB) surveillance and their films were confiscated. The Germans were very upset as their expedition has been sponsored by Stern Magazine of Germany and without the pictures, they could not have written any article. Bull had to again contact the Hon'ble Chief Minister who rang up the Hon'ble Prime Minister Mrs. Indira Gandhi directing the Army to get off their backs. Bull always amused by such incidents says, "Publicity is a double edged weapon indeed!"

TRISHUL SKI EXPEDITION - 1976

The Indus Boat Expedition earned Bull great amount of publicity that stamped him as the great adventurer and Maj HPS Ahluwalia who had climbed Everest asked him to lead the Trishul Ski Expedition. HPS had some contacts in Italy as he was treated after he had sustained severe spinal injuries. He made all the arrangements for collecting the finances, importing skis, arranging funds from Bharat Heavy Electricals Limited (BHEL) and State Bank of India. For Bull it was a dream come true. He had climbed Trishul in his first expedition in 1958 and ever since he had been thinking of skiing down the summit of Trishul (23,360 ft) as these slopes were very skiable.

At last came the the 17th of April - the day they were to leave New Delhi on their journey into the mountains. They as a team had looked forward to that day. And yet, despite months of hectic preparations, they had a nagging feeling that they had forgotten something. No expedition was worth its salt unless it had some last minute panics. The Trishul Ski Expedition was no exception. Some of their tents arrived just in the nick of time. They had to run around a bit for the oxygen cylinders. And at least one member found the last minute anti-cholera shot a bit painful.

But when they got into the bus to take them to the guest house of BHEL, in Haridwar, they were in high spirits though they were behind schedule. They were expected at the BHEL guest house at 8:00 pm but reached there only around midnight. The delay was partly due to road checks on the way. They felt sorry that they had kept the BHEL General Manager, Mr. Khosla, waiting for them at Haridwar. But they were given a warm welcome.

The next evening, they transferred to a mini bus. BR Sharma and BS Bajwa brought up their luggage in a truck. Bull always says that in any expedition, the transport member's job is an unenviable one. And these two did their job with a smile. After visiting Lakhsman Jhula, they went along the holy Ganga up to Deoprayag. It was exciting to see the Ganga emerge from the hills. They spent a night at Karanprayag where they slept in fresh linen and blankets. He mentioned the linen because during such adventurous expeditions, such luxuries are hard to come by.

Early next morning, they reached Joshimath in time for breakfast. Their Deputy Leader, Hukum Singh had gone ahead to arrange for their stay at Birla House; it was an ideal place for them as it had a lawns that served as good packing centre. Here they packed their luggage and porter loads. While the packing of loads and last minutes purchases were going on, Hukum Singh and Bull visited the villages of Rainee, Lata and Pang with a mission to engage as many porters as possible. It was quite amusing to see that wherever they stopped the locals asked them if they were from "Survey", as mostly the Survey of India officers visited that area for mapping purposes. For Bull, it was an experience to drive down the route from Joshimath to Lata as on his first Trishul expedition, he had walked right from Peepalkoti to Joshimath and then, to Lata village. It took them four days during that time, but this time they did it within an hour as the road had been developed.

In 1968, he had paid Rs 3 per day to a porter. But after eight years, they were paying Rs 15 a day per porter. Even the rations that they gave to the porters had been doubled – instead of half kg, they were demanding one kg. The worst was when having got all their demands; they put up yet another demand of going to Lata Kharak in two stages instead of one. That was not acceptable to the expedition party. The porters knew that they had the upper hand – with three other expeditions in the area at the same time, they felt that they could strike a bargain. Luckily for them, the District Magistrate intervened and they increased their wages slightly instead of making another staging camp.

On the very first day of the march from Lata to Lata Kharak, there was a steep climb of 5,000 feet that even experienced and tough men find difficult. Their caravan at that stage consisted of 70 porters and 70 goats. When they reached Lata Kharak, they found no water there. Melting of snow meant too much consumption of kerosene oil. As they suspected that the route was closed beyond Lata Kharak due to heavy snow on the cliffs; they decided to dismiss the porters for four days. They gave them food and wages and started work on opening the route through the cliffs. That was indeed one of the most difficult approach marches in the Himalayas.

The first day, they sent Gurcharan Singh and Lil Bahadur. They were followed on the second day by Bajwa and Major SS Singh. After the route was opened half way, they established camp to facilitate the work on the other side of the pass. At that stage, some Japanese joined them and they put a fixed rope on the entire 3 km dangerous route.

Their stay at Lata Kharak was in many ways helpful in the expedition. It gave them an excellent opportunity to acclimatize at 12,500 feet. It also gave them the time to try out their new skis that they had not used at all. On the 2nd of May they left Lata Camp for the Dharansi pass, that being on the southern slope, was not frozen. Then arose the problem of porters and they had to move in two shifts. The porters did a good job carrying the loads to the camp. In the meantime, a party was sent to Deodi to check whether the bridge over the Rishi Ganga was intact.

After spending the day at Dibrugata, they had a day at Deodi. The party reached the Base Camp (15,300 ft) on the 6th of May and the advance party on the 8th and no water was available there either. After considerable digging, they found a stream buried under six feet of snow and one foot of solid ice. They also faced the problem of feeding the goats. They tried to give them atta and rice but the goats were obviously on a non cereal diet. After two days of starvation, the goats consented to eat dal, but it had to be only urad ki dal.

Dorjee, BR Sharma, Vijay Kaul and Lil Bahadur then opened a route to Camp-I. On the 10th of May, they packed ferry loads for every member to carry. But when Bull reached the site, he found it too low for the camp and they shifted it to a height of 18,300 ft. The new site, apart from being higher gave them an excellent ski run down the ridge. Hukam Singh, Dorjee, Bhalla and Des Raj occupied Camp-II at the height of 21,100 ft. They stayed there for making the route up to the Camp-II as the previous day; the snow storm obliterated whatever tracks they had made.

When Bull reached Camp-I on the 19th of May with Zargar, he was very upset to find that the second summit party had left without his approval. It was a bad precedent, to say the least. But in a way he was glad because in the overall interest of the team perhaps it was a good decision as the weather was favourable. He felt sorry for Zargar. He was so keen to ski down from the summit; how could he go then? There was no support party and there was no member to accompany him. Zargar was furious. He roundly abused everyone in the second summit party and said, "How can people be so selfish?" Bull tried to console him by saying that he would keep his promise that any member of the team who was fit and desirous of going to the summit would get a chance before they called expedition off. Zargar laughed sarcastically. Bull said that if he was so keen he would act as a support party and take him to the top. Zargar did not make up his mind for quite some time. He was sorely disappointed and was obsessed with negativity.

Zargar, Sampat, Jabbara and Bull were the only four people in the mess tent and the camp looked quiet and deserted. Jabbara was feeling sick but he had stayed on to see all the summit parties through. For a man who had never climbed upto 18,000 ft, to stay there for seven days and prepare good food, showed a lot of guts. He reflected on those days at Lata Kharak when Jabbara complained of piles and wanted to go back. Bull had then asked the doctor to attend to him and he said that as Jabbara had waited for such a long time, there could be no harm in his staying another few days and he assured Bull that his sickness would not be aggravated because of the high altitude. Bull Kumar had kept Jabbara forcibly against his conscience for the sake of the team and there he was proving to the world that his decision was justified. That afternoon he produced such an excellent roasted mutton of a quality that Bull had seldom tasted before even in a 5 star hotel! There was little doubt that Jabbara, with his great skill at improvisation, was an ideal cook for the expeditions.

The afternoon was not entirely pleasant and Bull kept thinking of Zargar all the time. At 4 o'clock in the evening he came to him and said, `Sir, I will take you at your word. I would like to go to the summit'. The die had been cast; the ball was now in Bull's court. He had made a promise that he had to keep, even though it might be hard to do so. Since Bull had lost his toes, he had been advised by doctors not to go beyond 7,000 ft. But Bull had been indulging in all sorts of base camp climbing.

Bull had led many expeditions after he had lost his toes, but in all of them he was an administrative head rather than a lead climber. He had been told by the doctors that if he exposed his feet to cold again he might lose them. He had promised his friends and his family that he would not be going very high on the mountains; at the most he would climb up to the Base Camp or Camp-I. But there he was forced to make an exception because he felt committed to his team. Perhaps Zargar saw some tension in Bull's face, and said, "It is not a compulsion, Sir. If you don't feel like it, it is okay by me". Though Bull told him that he would keep his promise, he was in fact feeling miserable. After having lost his toes he had never attempted a summit or gone to such heights. This would be the first time that he would be taking such a risk and he wondered if it was worthwhile. He told Zargar, "We shall go tomorrow morning, but on the condition that whenever my feet get cold we will return". He replied, "That is okay Sir".

The next morning Bull got ready, fixing his electrically heated socks. These socks worked on torch cells attached to the upper portion and there was a warming portion near the toes working on the same principle as an electric blanket. He tried both pairs that worked well. But every hour a battery had to be changed. He had to carry a minimum of twenty cells with him- almost two kgs of extra weight, and yet he was not sure if that would help in that extreme climate. With only

Sampat and Jabbara to bid them farewell, Zargar and Bull, along with two porters Surinder Singh and Meharban Singh, started for Camp- II.

Though the porters were saying that they would support them to the summit, Bull knew that they had neither the intention nor the capacity to get to the top. In fact even though they were carrying fewer loads than them, they had difficulty in getting to Camp-II. After about an hour's journey they saw the second summit party returning. In spite of his grievance against them Bull was genuinely happy that they had made it and come down successfully.

The moment they got to Camp-II, they rushed inside the tent. It was full of snow. Someone from the earlier summit parties had left the tent open. However, in that terrible storm just to be inside the tents was a great relief. It would have been difficult to burn the stove but the little amount of Butane gas that they had with them came in handy. They had one burner with three refills. It was a very neat arrangement. One could fix the refill in the frame and then puncture it by rotating the burner. But one had to be very careful while doing that. Sometimes the socket did not fit properly and the gas would come out at a tremendous pressure. As the escaping gas expanded it became cold that could cause instantaneous frost-bite to the fingers or any part of the body that came in contact with that.

In fact that was about to happen when Bull fitted the socket but he had taken the precaution of putting on his gloves before he handled the stove. Then it struck them that very soon the tent would be full of fumes and if a match was lighted the entire tent might get engulfed in fire. He told Zargar to throw away the cylinder quickly, that he did, and after ten minutes of hissing and shrieking the stove was calm again. Hesitantly, as if he was touching a live electric wire and with his hands and face covered, Zargar extricated the burner from the empty cylinder. Luckily, they still had two spare cylinders and they tried their luck again with the second one. This time Bull was very careful.

It took them almost half an hour sitting around the fire before they could feel any sensation in their bodies again. If they had remained in that blizzard for another hour or so could well have been fatal! They removed their boots and warmed their feet. They warmed their hands and put them on their cheeks, ears and noses. No porter turned up for fifteen minutes so that they had the fire to themselves. They then heard Surinder Singh shouting "*Mar gaya, mar gaya, aaj to mara gaya ...*" (I am dead, I am dead, today I am dead) from outside. They asked Surinder to get into the tent. He was blue all over. They let him use the stove exclusively for about ten minutes and he revived. But where was Meharban Singh? Somebody had to go out and search for him; otherwise they would have a case of hypothermia on their hands.

Obviously that meant a lot of inconvenience to put on the cold boots again and go out into the blizzard. But that had to be done; and in such circumstances the leader would feel very guilty if he had to send his companions out. Therefore, Bull wore his boots again and the moment he got out of his tent he saw Meharban Singh lying completely numbed by the cold. He had dropped just a couple of feet away from the tent. Bull helped him to get up and brought him inside the tent. He himself sat out because there was not enough room in the tent for four people. He utilized that time clearing snow and cleaning dirty utensils that the second summit party had left behind.

Since he was also getting cold, they shifted the porters to the second tent and lit the stove for them. But they refused to cook their own tea or eat anything. They just wanted to go to sleep - a natural phenomenon at high altitude when one was tired and cold. Bull knew, however, that unless they ate something they would hardly be able to do any work the next day. They might not even be able to go down to Camp-I. Instead of being a help they had now become a burden. He doubted if ever before there had been a case at that altitude where the leader and members of a mountaineering team had prepared tea and inflated mattresses for the comfort of their porters, as they did.

But porters were not prepared to eat anything, not even the 'parathas" that they had brought from the Camp-I, nor the tasty roasted meat that Jabbara had prepared for them. They only wanted to vomit and sleep. They warmed some "parathas" for them and when Surinder Singh heard that they also had chilly pickles—red chilly mixed with garlic—he consumed a couple of "parathas" with great relish. But Meharban Singh would not touch them and Bull realised that he was not going to be of any further help to them whatsoever.

They heated up two tins of tomato juice, one for themselves and the other for the porters. Ashraf now became the cook of their tent. They enjoyed "parathas", juice, and soup followed by Bournvita in condensed milk. They used half the tent as a kitchen and thought that they would be able to sleep comfortably in the other half. While Ashraf took a sleeping tablet and dozed off, Bull spent some time in warming up his boots. He had to be extremely careful about his feet, otherwise the expedition could end in tragedy. He had undertaken that ascent merely to give a fair chance to all members of the team to get to the top.

For the next two hours he kept on warming his boots, dried his socks and then tried to sleep. But how could two men of their build sleep on one air mattress? For another two hours thereafter, he kept sitting, not having the heart to wake up Zargar. Only when Zargar got up at about 3 o'clock they decided to clean the kitchen and put another mattress so that both of them could sleep. But at that time the wind was blowing hard and too much snow had piled up on their tent that could collapse.

Though they were tapping the tent roof from inside to dislodge the snow, they had given up all hopes of reaching the summit. They had been warned that there was a second disturbance somewhere in the Middle East that might or might not affect them. Bull was sure that, that had affected them and they resigned themselves to their fate. He was happy in a way that he did not have to climb to the summit, because he was not at all sure about his feet.

At 6:30 am Zargar got out of the tent and exclaimed, 'Sir, we have been deceived'. "What has happened?" Bull asked. He exclaimed, "There are no clouds anywhere. The weather is excellent. It is only the strong wind that is blowing the snow around our tents!"

Bull thought it was a snow storm. However, they decided to go up even though they were three hours late as compared to the first summit party. They shouted across their tents for the porters. There was no response. They tried to induce them by promising to give them higher wages but it had no effect.

When he went inside their tent Bull saw that they were lying cosily in their sleeping bags. He was quite sure that they were not sleeping because only a superman could sleep soundly at such an altitude. So he shook Surinder Singh and said, "What is that nonsense? If you do not want to go up, say so" Surinder guiltily got up, but Meharban Singh did not.

It was 7 o'clock before they left Camp-II. Unfortunately because of their unpreparedness to leave, they had not yet had any breakfast. They had half-a-tin of juice and left for the summit. Luckily, they also carried a small bag of dry fruits. Then they faced the difficulty of tying the skis. They tried all possible methods but every fifteen yards or so they kept becoming loose; till they got fed up and adopted the most unorthodox method of tying the skis straight to the back and wearing the rucksack over them.

The first thousand feet climb was steep and there was hardly about a foot of snow that had to be beaten. Surinder went up for about two hundred feet and then sat down saying that he could not make it. With great difficulty by putting him at the rear they made him go up another two hundred feet. Again he sat down and said that that was the highest he could go. No amount of persuasion, entreaty or promise of additional reward was of any avail. Both of them had to then share his load and they started again. Being the senior member it was Bull's lot to take the lead and it was tiring indeed. But he was very keen that Zargar should get the same help that other people had got or as much of it as possible to enable him to ski down.

They now stopped and took some raisins. That was their breakfast and lunch combined. The wind was strong and the wind-driven snow was playing havoc with them. They waited at the top of the ridge till Surinder got into the tent at Camp-II and they started again. After about 100 ft of straight climb, they continued climbing on a gradual slope, at times steep and with many humps. These humps could be very deceptive. One felt that one had reached the summit but when one got to the top, one would dishearteningly see that there was another huge slope ahead of one. There was one particular rounded hill that looked like the summit but turned out to be another glaciated slope that extended towards the real summit.

They carried on and on and Bull was leading and Zargar was following him. Every time he looked at Zargar he said, "We shall make it today. We must make it today." Then the clouds started coming from the west. They tried to quicken their pace but it was of no use. They could not accelerate the pace as they had been climbing for five hours and were fatigued. Bull admitted that it was a slow pace for the reason that not only were they carrying heavier loads but there was nobody else to beat the track for them as had been done for the earlier two summit parties.

When they were just about forty feet below the summit, Zargar said, "I think we should not go any further as we may not be able to get down skiing because of poor visibility". Bull was happy that his feet were standing the cold so well, and as he had given his word to Zargar, he was prepared to climb as high as he wanted. The peak was then only half-an-hour's distance at the most, but they abandoned the climb and started putting on their skis. They were extremely careful to see that the skis did not roll down. Zargar helped him to put on his bindings that he found, under those conditions of extreme cold, worked very well. Later on he stepped by his side to help him. He yearned for some liquid refreshment.

They were terribly dehydrated and by then even the raisins had been consumed. His camera without its case had become a real problem. He had to put it in the rucksack; otherwise it would have been ruined. It was not possible for him to take that out more often under the prevailing conditions.

It was 3.45 p.m. when they started their descent. Down at the Base Camp and at Camp-I, the others were terribly worried about them. They could not figure out as to why one man had come down so quickly to Camp-II, and why the party that had taken almost twice as much time as the others, was still not visible to them. If they had not appeared for another hour or so, perhaps a message would have to go to Delhi that they were missing. Some of the members were afraid that some tragedy had befallen them otherwise why should they take so long?

The moment they put on their skis and started coming down they felt an ecstasy they had never experienced before. There is something grand about skiing. However tired you may be, you always want to have a last round. Although they were greatly fatigued and cold, they still found skiing a great pleasure. By then there was only a little clearing left on their descent route to show them the general direction.

While skiing down, Bull took a little traverse to the route where he found snow conditions very smooth. He hurried his descent. The turns were so easy that it looked as if the whole slope had been packed for them by a snow grooming machine; but that pleasure was short lived. Very soon when he looked down he found a drop in height of six thousand feet or more. He was very surprised. He had studied the map so carefully and had never thought the slopes on that side were so dangerous.

He remembered the discussions he had with Major Hukam Singh and Major SS Singh on the fixing of flags to mark the dangerous areas and they were under the impression that the northern slopes were more treacherous than the southern. There they found to their dismay that the contrary was true.

With that realization came the shock that Bull had committed the worst mistake of his life. He was skiing on a huge cornice that was protruding for about thirty feet in the air. His heart almost missed a beat; the cornice could break at any time and that would be the end of the skier. He had a second look to make sure that he was right and that was again most frightening because he realized that wind was blowing towards the cornice!

He was then certain that the nice smooth area that he had decided to ski on was nothing more than a death trap. Even with his sluggish mind - as at that altitude his mental efficiency was halved due to lack of oxygen - he knew that he could not put too much pressure on his skis for turns as the jerk might break the cornice. But there was another factor that had to be considered.

If he had to take a smooth turn, he would have to go much further outwards towards the edge of the corner, where the cornice became thinner and thinner. He felt himself completely trapped. With his heart in his mouth, a chill in his stomach, sweat on his forehead and weakness in his legs, he leaned slightly inside and opened his outer ski. To stem without applying any heel pressure whatsoever, he turned to the left with a longer stem swing, hoping that the cornice would bear his weight. He took just a couple of seconds, but that felt like eons!

Many ideas flashed through his mind in those moments of crisis. Why did he come on the expedition? Why did he come in the summit party? If misfortune befell him what would happen to his family, so wholly dependent upon him? Was it fair or reasonable to have taken such a risk? And finally he thought of the comrades who perhaps would rescue him or might themselves become victims of the mountains - precious lives sacrificed at the altar of nature and bold (mis)adventure!

He brought his skis to a sudden halt, once he was sure that he was on solid mother earth. The fright seemed to have taken all the energy away from his legs, his knees sagged and he found his lungs bursting. He had not sufficiently recouped from the shock when he looked up. Involuntarily a cry escaped his lips, "Oh, no!" His knees found strength again, he got up and his bursting lungs were shouting madly, "Zargar! Zargar! Cornice, come back!!"

As his mouth opened he looked at Zargar following his tracks and he knew that he was running a far greater risk of going down than him, because after bearing his weight the cornice had become weaker. He clenched his sticks and shouted again and again! In between the shouts he would throw in a prayer, "Oh! God, please help us." The situation seemed to be completely out of control.

A thought came to his mind. Rafiq Burza, a classmate and colleague of Ashraf Zargar while under training as a mountain guide, had dropped dead from a rock right in front of their eyes. He had been married only for four months. He remembered how upset Zargar was by that tragedy. He had not eaten anything for days and had cried like a child, and then he himself was going to meet a similar fate. Bull cursed himself for having brought Zargar up towards the summit.

He did not know how long he kept gazing and shouting and whether that had any effect on Zargar but the next moment when realisation came to him, he found Zargar standing near him. "Did you see that drop?" Zargar was shouting against the wind. Bull wondered if he realised what the drop meant. "That was a cornice, my friend, and you just had your closest brush with death," he said.

He could see Zargar's face going pale and then he shrugged his shoulders "Well, it's all over. Let us hurry up, clouds are closing in." Bull looked down and said, "My God!"

They needed a lot of luck to get out of the mist and the mess into which they had landed themselves. There was a narrow ridge in the front like a culvert. They must get over it while they still could see something.

"Let us be careful, we cannot afford a broken leg now," Bull shouted.

Down they went skiing again. Straining their eyes to find the marked flags through the mist and fog in the strong blizzard blowing snow into their necks, noses, ears and eyes, they skied down for survival. That was definitely not pleasure skiing.

"Will we make it?" Zargar asked. "Of course," replied Bull.

He had spoken in a confident tone that uplifted Zargar's morale. Only he could feel the emptiness of those words because he knew that if they skied down left or right of the bottle neck, perhaps they would never be able to get to the Base Camp. Then they saw a little opening just above the bottleneck. He thanked God for His mercy and let go their skis almost at full speed. He knew that once they were on a broader slope across that "culvert," they could only ski down towards Trishul glacier and towards the Base Camp.

After crossing that narrow bit they halted and had the last few raisins still left with them. The snow became a little softer but skiing was still amazingly easy. Their skis were turning as if by remote control because, as far as he could think then, there was hardly any strength left in their knees to do intricate movements of parallel turns. They came down and down, turning merely with inward leans as they had no energy to push their legs out as was done in Austria and to use all the usual manoevres.

In the distance he could see some broken lines in the whiteness and he knew they were nearing Camp-II. There was a huge crevasse on the left of the camp that he had noted while going up and was aware of its danger. But before he could see the crevasse once again a mass of clouds came up as if a giant Yeti was blowing out smoke. The camp and the area around got completely covered. They decided to wait for a while till the clouds lifted again.

After waiting for five minutes the clouds did not show any inclination to move away. So they started skiing again, carefully. Zargar was ahead of him. He could just see the outline of his figure, when he heard the words "We are at Camp-II". New life seemed to have been infused into him. The prospect of hot tea and coffee and some food after nine hours of hard work was most welcome.

As soon as he reached Camp-II, Bull found Zargar abusing like a mad man. "These bloody porters have ditched us again!"

"What has happened?" Bull asked.

"There is nobody in the tent!" Zargar replied.

All their hopes of having a hot cup of coffee with some food faded. They had to make a choice to spend an hour or so trying to light up the stove and eat something or suffer the pangs of hunger, thirst and cold for another hour and be comfortable in Camp-I, where everybody would be waiting for them. They decided on the latter course and started for Camp-I. But by then the snow conditions had become very sluggish.

There was wet heavy snow and fast short turns seemed impossible. Instead they started taking long traverses. Even then, many a time they lost balance when the skis got stuck. To negotiate that sort of snow one really need some strength in one's legs while they had none. And then with their energy almost completely drained and their spirits flagging, with a sigh of relief, they spotted Camp-I where their comrades were all waiting for them.

In Camp-I every one was thrilled to see them back because there was a time when they had given up all hopes of their survival, especially when even the porters had deserted Camp-II and had gone down. They had seen them crossing the false ridge and should have expected them back within about three hours but they had actually spent five hours. The weather was bad, it was getting dark. They decided to spend the night at Camp-I. They were completely dehydrated and drank a lot of juice and beverages. Both were happy to be back at Camp-I because then Bull knew that the expedition was finally over and they were all safe. The next morning they bid "au revoir" to Camp- I and skied down.

Their run from Camp-I to Base Camp was perhaps the most enjoyable run of his life and only for that moment, the entire expedition seemed worthwhile. Sampat followed them on foot with his camera. Unfortunately, he had spent most of his film already and had only a hundred feet left. They did not take the normal route taken by the other skiers but went below the ice-fall on their right because the hanging glacier up there looked pretty sound and the slopes below it were fantastic.

Nearing the Trishul glacier he saw Zargar come to a halt. And when he closed in, he said,

"I almost went in for water skiing."

Bull looked down and realized how right Zargar was. There were small puddles of water covered with a thin sheet of ice. They skirted around them and came down and found that the last 200 yards were nothing less than a stone slalom. The rocks were already showing their black faces and very carefully they came down, trying to avoid even a scratch on their skis. The entire ski down 3,000 ft had only taken them 15 minutes but it was one of the most enjoyable parts of their adventure.

They were having breakfast on the 21st of May when they suddenly heard the whirring of a helicopter. They had sent a crash message to Major Ahluwalia in Delhi to send a chopper to the Base Camp to evacuate Mohd Bhatt. It flew in without any prior intimation. They gave a smoke signal to the pilot. He flew over the whole area again and again but could not find any suitable spot to land. He must have left in disgust but before doing so he was good enough to point out to them the place that could be developed into a makeshift helipad. That was the place where they had pitched their mess tent.

They shifted that tent and the others elsewhere, shoveled the snow and levelled the ground and marked the alphabet "H" with atta. Under normal conditions the helipad must be 30 yards by 30 yards, completely plain and without any obstruction within 200 yards with a good approach and take-off and not be at a height above 10,000 ft. But all these conditions were impossible for them to meet. The maximum sized helipad that they could make was 30 ft by 30 ft and it was at a height of 15,000 ft.

They were not sure if the helicopter would come again, so they sent the following crash message to Major Ahluwalia: Ski Trishul (.) For Major Ahluwalia Delhi (.) Copter came unexpectedly and flew over Base Camp for about an hour around 0900 hrs today (.) Pilot tried best to land (.) Members available at Base Camp immediately marked a helipad and gave smoke signal but helicopter could not land since area was covered with snow (.) Finally pilot tried to land near camp but could not do so due to tents (.) With the indication that pilot intended to land at that site all tents have been removed from the area and it has been cleared of stones (.) Request convey that info to concerned IAF authorities (.) in case copter cannot land at Base Camp will prepare helipad at Deotoli as already intimated (.) request intimate expected time of landing of helicopter (.)

Ski Trishul (.) From Leader to Major Ahluwalia Delhi (.) Further to my sig No Ext-59 of date reg air evacuation (.) As a result of detailed reccee of that area only possible helipad site available is at DEOTOLI (.) Map ref already intimated (.) Details of helipad altitude about 12,000 ft dimension 30x 30 yds (.) Surface free from obstructions (.) That helipad is not tested (.) Helipad site can be made suitable for landing within 18 hrs on receipt of intimation (.) Smoke signal can be arranged for indication of helipad (.) General direction from JSM

(Joshimath) along Rishi Ganga in Rishi GORGE and Trishul bank (.) Team scheduled to leave base camp for JSM on 23 May (.) Request expedite evacuation (.)

The IAF helicopter came again on the 23rd of May in the morning. The young pilot deserved great credit for making a perfect landing on a very small strip at such a high altitude! Hukam Singh accompanied Mohd Bhatt and admitted him to the Military Hospital at Joshimath. The rest of the team left in the afternoon. When they came to the dangerous Dharasi cliffs, they found to their consternation that the 3 km long rope fixed to make the route safe had been removed by some irate porters, probably to tie their cattle. To add to their misery the weather turned extremely foul. Two of their porters fell down and were injured. Dr. Gupta managed to bring them back to safety with the help of a rescue party and gave them tea.

The team had still to perform another life-saving act before reaching the road-head. Midway between Dharasi cliffs and Lata Kharak they found a few unknown porters shouting and gesticulating in panic. They told them that an officer of the Geological Survey of India from Lucknow had fallen down 1500 ft from the cliffs. The porters had stood by helplessly from 11am to 4 p.m. and left the officer at the mercy of God. Although Bull's team members were very tired after a long march that day, they immediately put on their skis and rushed to the place where the officer, Mr. NK Malhotra, was lying unconscious. Had he not been pulled out to safety, he would have frozen to death during the night. He was carried by the members to the Lata Kharak camp and put in a tent next to the kitchen to keep him warm. Dr. Gupta made indeed supreme and successful effort to revive him back to consciousness and health.

KANCHENJUNGA - 1977

After returning from Trishul, Bull was in Joshimath. They had successfully skied down from Trishul - the highest summit ever to be skied down. Every success had left a void and emptiness in him. It was with that feeling that he reached home where he found an invitation to dinner, being hosted in honour of General TN Raina, MVC., The Chief of the Army Staff, Brigadier Chowdhary, the host had added in the postscript 'You must come'. There was no question of declining that invitation, but he could not attend it, as he had no party clothes. To go in front of the General in his jeans would have been unforgivable and how he wished that he was then up on the mountain. For the next one hour he sat at the telephone trying to find out the heights and other statistics of various officers in the station. Luckily he found some one of his size and borrowed a lounge suit to attend the party.

General Raina had just read the *Hindustan Times* that covered his story about Trishul Ski Expedition. He told Bull that he had always found that these adventure expedition leaders and summiteers were from the Army, but the credit always went to the bureaucrats and the politicians. With his permission, Bull asked him a question as to who climbed the Everest in 1953. His quick reply was the British Expedition. Bull told him that the British only organized and paid for the expedition, the people who climbed the Everest were Edmund Hillary from New Zealand and Tenzing Norgay from India. He got the message and he asked Bull to arrange for some big expedition for the Army. As he was also the Colonel of the Kumaon Regiment, Bull could not say no and later gave a proposal that Army would climb Kanchenjunga from the North East Spur.

Why Kanchenjunga? In 1953, after the climb of the Everest, Lord Hunt was asked, "What's Next?" "Kanchenjunga," he had replied. The greatest feat in mountaineering with technical climbing problems and objective dangers greater than the Everest was Kanchenjunga. It was only after 50 years struggle that that mountain gained the reputation of being not only the most beautiful but also the most dangerous mountain in the World.

It was climbed in 1955 by a British team led by Charles Evans. But the reputation of Kanchenjunga had been established more for its treacherous Eastern approach that had defeated two Bavarian Expeditions, led by Paul Bauer and composed of the most brilliant German climbers of the time. It is said that those Expeditions fought their way, by foot, through the towers and mushrooms of snow and ice that crown the ridge by an operation of carving and tunnelling, quite unheard of at that high altitude. The Alpine Journal that rarely indulges in superlatives described the Bavarian attempts on Kanchenjunga as "feats without parallel, perhaps in all the annals of mountaineering".

As far as Bull was concerned, apart from the fact that climbing Kanchenjunga from the East was the greatest challenge available to any mountaineer in the world, it happened to be his country's highest mountain. It was but appropriate that Indian mountaineers made serious attempts to climb that peak. As head of the Indian Institute of Skiing and Mountaineering, Gulmarg, it was not possible to organize the expedition. So he decided to revert back to the Army.

It was quite obvious that the expedition needed tough and experienced climbers. The selection had to be broad based. They decided to hold a selection camp in Nov 1976 for that expedition. 35 climbers came to show their mettle; and a selection committee consisting of heads of Mountaineering Institutes in India and some Everesters, selected a team of 16 members. Some of the better known names including Major Prem Chand who was perhaps the only mountaineer who had climbed both the Nanda Devis; Major SS Singh who climbed Nun and skied down from Trishul; Major Pushkar Chand who had to his credit Umba peaks; Captain KI Kumar who has been a member of Nanda Devi, Saser Kangri and Brahma expeditions and had climbed Leopargial Nima Dorjee and Norbu and also had made the first ascent of Sickle Moon, the highest Mountain of Kistwar; and Nirmal Singh, an Instructor of the Nehru Institute of Mountaineering who had been up Nanda Devi.

In 1970's the foreign exchange position of India was very bad. In fact the time had come India had to mortgage gold to get dollars to buy oil. Under those conditions to get any foreign exchange for a sport like mountaineering was almost impossible. However, they were lucky to get oxygen bottles and high altitudes mountaineering equipment in Kathmandu purchased with Indian currency. Beside these, there were major problems with the oxygen bottles as they did not fit on the regulators they had, but Bull approached an Electrical Mechanical Engineer (EME.) of the Indian Army who came to their rescue and made suitable adapters. Most of the equipment used for the Army High Altitude with no doubt was very durable and warm, but also very heavy for higher altitudes. Their two men tents weighed 12 kgs and were made of white cloth for camouflage. They had added additional coloured stripes for visibility and photography.

The German expeditions that attempted that route in 1929 and 1931 had done so in the post mountaineering season. After consulting the Indian Metrological Department, Bull decided to go in the pre mountaineering season.

As there was a shortage of porters in that area, Bull requested the Army to help them out. Accordingly, they recruited porters from Gangtok and surrounding areas. Even though they had given them full warm clothing as was given to the high altitude soldiers, with the first snow fall after two days, the porters deserted and left them in the lurch not even taking their wages. However, Bull had made provisions for such an eventuality and left plenty of rations at Bagdogra packed in 25 kgs bundles that were dropped by the Indian Air Force (IAF) at their Base Camp. Also, they had many ferries to Base Camp to carry their technical and mountaineering equipment.

He knew it would take them almost a month to get their stores across to Base Camp. Therefore, they sent forward small parties to start work on the mountain. The first party under Major Pushkar Chand left Lachen on the 17th of March and established Base Camp on the 21st of March 1977 in the Green Lake area. The name was misleading as there was in actuality merely a brown muddy water pond. Perhaps, after the monsoons the pond enlarged itself to become a lake!

Theirs was the earliest entry in that area and even to get to the Base Camp, they had to make through very deep snow and "going" for the porters was extremely difficult as the Zemu valley, especially during the first two marches was very narrow and steep. Two huge seasonal avalanches had blocked the tracks and they had to find their way through. But after the Rest Camp (third Stage), the valley became broad and there was hardly any snow.

Kanchenjunga is a religious mountain and like Nanda Devi in Garhwal Himalayas and Chomolhari in Bhutan, people worship it. Therefore, they decided not to climb right to the summit but to stay six feet below as was done by the British Expedition in 1955. However, the following press statement was issued from Rhumetek Monastery: 'Kanchenjunga is angry'- by our correspondent: Gangtok March 31: "Monastic circles here have reported what is described as an unusual phenomenon connected with Kanchenjunga.

Lama Karma Gyalpo, who belongs to the old Rhumetek Monastery, said here yesterday that he had received a report that a big explosion was heard from the direction of Kanchenjunga on March 27. Heavy landslides, perhaps avalanches, followed it the next day. The Lama said the phenomenon had affected the nearby Tongshiong Glacier. According to monastic circles, such an incident had never occurred in living memory. Scores of coolies engaged on road construction work near the green Lake had suddenly fallen sick.

Reports from Mangan said that tens of thousands of fish had been sweeping down the Rakhel Chu and Talung Chu Rivers to Sangkhalang during the past two days. The Rakhel Chu originates from the Tongshiong Glacier. According to karma Gayampo Lama, it is considered by the local people as a manifestation of the wrath of the Gods at the attempts being made by an expedition to scale the mountain, that is the ' protecting deity' of Sikkim and hence sacred. Prayers are being offered at Lachung Gompa to avert ominous portends".

Bull was not sure what affect it had on the climbers but surely many porters deserted them. On the 28th of March 1977, Major Prem Chand opened the route to the Advance Base Camp that was established on the moraine of Zemu Glacier, a little ahead of its confluence with the Twins Glacier. It was only when one reached Advance Base Camp that one realized that one was approaching one of the biggest mountains of the world. There, one could see encircled by the mighty Kanchenjunga rising almost 10,000 vertically and every time one looked at that, ones head hit back the rucksack frame. Towards the North were the brown mountains like those of Ladakh, full of glaciers in that bowl of not more than 26 square km and over 50 small and big glaciers. They were like frozen water falls. They roared with avalanches and looked very aggressive but once they fell into the great Zemu Glacier they were like mountain streams crawling into the mighty ocean. The Zemu Glacier was reported as the slowest and the most placid glacier of the Himalayas.

Having read all the known books, the accounts of almost all the expeditions on Kanchenjunga Mountain, the maps in details, pondered over the aerial photographs for hours and undertaken helicopter reconnaissance of the area, Bull came to one conclusion; that the North east Spur Route that was taken by the Bavarian Expedition and considered one of the most difficult and dangerous route, compared to those of the other similar mountains, was still the best one to the mountain from the east. They had no choice but to fight the dangers and try to break through the defences of Kanchenjunga as the Germans did.

Some foreign mountaineers had suggested a route from the north ridge from the Col between twin peaks and Kanchenjunga and some had even suggested one from Zemu Gap to South Kanchenjunga leading onto the main peak. Bull was of the view that human beings as they were, one day surely would climb those routes also.

After the Advance Base Camp was established they negotiated the icefall. Though a small one, (only 10,000 ft) that created quite a few problems in the beginning. One of the members Naik ND Sharma fell into a crevasse and was saved by timely action of his rope mates. On another occasion Captain Cruz and his party fixed a rope and climbed up a portion of the ice fall to recce the higher approaches but on the return journey found that the fixed rope had been swept away by an ice avalanche. On the 1st of April 1977 Maj Prem Chand's party successfully crossed the icefall and established Camp-I at the height of 5,720 meters. The route between Camp-I and II lay through steep snow and sheer rock faces. Dangerous going but Major Pushkar Chand, Nirmal Singh and Gurcharan Singh made it comparatively safe by fixing the entire route by rope. Safe was a comparative term as they would see later.

While taking a ferry to Camp-II Naik ND Sharma found the old camp site of the Germans under the Eagle's nest. Almost 50 years old remnants of the Bavarian Expedition were found there. Those included crampons, spoons and snow shoes.

On the 10th of April Kiran's party joined to find out how difficult and dangerous the traverse route would be. Going from Camp-I was a flat crevassed glacier for ten hundred yards up with steep faces of mixed ice and rocks like the top of a knife edged spur. Kiran had cut steps, fixed rope; and slippery rocks traverse vanished, when Bull saw what lay ahead of them. It was a steep arête of ice rising for 2,000 feet that had fluted a 4,000 feet drop on the Northern side and a cornice with a 2,000 feet drop on the southern side. It was no less dangerous that the horizontal traverse that they had done earlier. It seemed that objective hazards were going to haunt them right up to the summit.

On the 11th of April when Captain Kumar's party had nearly finished the traverse they found Major Prem Chand coming down the arête. He had not only found a successful route to the end of the traverse but also had gone a little higher. Bull had discussed the route taken by him. He had seen both the routes -- the first one up to the Camp-II that followed the general line of the German route that he had then opened. According to him, the new route was much safer and shorter and therefore, they decided to abandon the traverse.

Captain Kiran was disappointed that he could not reach the end of the traverse. It had meant a few more hours of work. Keeping in view the larger picture and common benefit, he reconciled.

The 12th of April dawned bright. They all had a little breakfast and descended from Camp-II. Company Havaldar Major (CHM) Norbu was leading, followed by Havaldar Sukhvinder, Kiran, Kura Ram and Bull. They were all coming down independently on the fixed rope. The distance between them depended upon the terrain. Whenever, there was a danger of stones over 100 yards, climbers had to go one by one. Kumar was at the most precarious slope when he heard shouts for the very steep gully becoming smaller and smaller each moment. And then there was another shout 'Kura Ram help'. Kumar asked Kura Ram who was behind him to drop his rucksack there and rush to help.

Bull was little scared, when he fortook the safety of the fixed rope in order to cross. But he was very sure footed. After Kura Ram had left, with great anxiety he crossed the difficult patches and arrived at the site of the accident. There stood Kiran and Kura Ram over the dead body of Sukhvinder Singh. It seems while rappelling down, Sukhvinder Singh slipped, lost control, hit against a rock, broke his neck and died on the spot. Forty five years ago, a German and a Sherpa had fallen off that very cliff and died.

For half an hour they sat looking at one another, hoping against hope, feeling Sukhvinder Singh's pulse again and again. It was strange but in that state of shock; at times any one's pulse could appear to be that of Sukhvinder Singh's and a ray of hope shined momentarily. The question now arose as to how to take the body down. It was quite obvious that it would be impossible for only three of them to carry the body down the precarious slopes. Normally, expeditions buried the dead near the site of the accident, but they were determined to bring the body down and give it a decent funeral. They shouted for help to Camp-I. After considerable shouting they saw a line of people moving up from Camp-I.

That party from Camp-I came up in a remarkably short time. They started making a stretcher out of ruck sack frames. It was a Herculean task to bring the body down the steep slopes. They reached Camp-I just before it got dark.

Bull contacted the Base Camp on the wireless set and told Major SS Singh to send the following message to the Army Headquarters "Regret to announce the death of our Sukhvinder Singh in a climbing accident while he was coming from Camp II to Camp I. Body being taken down to Base Camp for cremation."

The next morning, they attached a jerrican below their sledge to help slide the stretcher more easily. It was a quite problematic to bring the body down the ice fall and quite a few people got sprains and back aches because of the awkward positions in which they had to walk, carrying the body. At the end of the ice fall, the body was re-tied to the stretcher firmly. While they were doing so, a massive shower of rocks came like bullets from the top where the 1931 German Expedition had buried the body of a German climber. Bull had never been as anxious in his life as he was then. Twenty meters behind the funeral procession, he sat down to pray for the safety of the pall bearers. He did not stir from there until the body had cleared the dangerous area, then thanked God for His mercy and joined the "cremation caravan".

As arranged, Sherpas from the Base Camp had come up to relieve their party and carry the body down to the Advance Base Camp. Walking on the glaciers moraine with the body was a difficult task. They had to spend a night at the Advance Base Camp and the next day porters came from the Base Camp came up and relieved the Sherpas in carrying the body to the Base Camp. For the next three days it snowed heavily and they could cremate the body only on the fourth day.

All those days, Bull had been trying to think of what should be done next. The morale of the team was very low. The task ahead was one of the most dangerous in mountaineering. Time was running out. He was certain that they had to do their best to fulfill their task in the light of the fact that one of their comrades had died. It was the turn of Major Pushkar Chand's rope to start where they had left off, but that could not be, as Pushkar himself was ill and Gurcharan Singh had a terrible backache. Major Prem Chand volunteered to go back to the ridge. -Bull decided to accompany him to Camp-I. Thanks to the discipline and training that the Army inculcates in its personnel, the team rallied round him to carry out their mission.

The work on the ridge commenced again with Prem Chand in the lead. At that stage, one of their ace climbers, N.D. Sherpa had to be taken off the mountain as he was suffering from "high altitude piles". However, Prem Chand kept up the work with the help of Anchok and Dorjee at appreciable speed. Dorjee, too, had to come down because of "high altitude sickness" but Anchok and Prem Chand kept going slowly forward.

The route from Camp-I to Camp-II stretched over half a km of ice plateau, almost level, up to the bottom of the North East spur. Then that went over a huge snow and ice debris where all the avalanches started. Frequently, the avalanches from the top obliterated their track through the ice cone. There was only one saving factor, the slopes were so steep that any snow, that fell on them, kept sliding down, causing small avalanches. Many a time climbers came under these avalanches. After the avalanche passed the climbers shrugged off the snow and started climbing again.

After the cone of snow debris, the route became a rocky platform under a rocky gendarme that the Germans named Eagle's Nest: here they found three crampons left by the Germans in 1931. From Eagle's Nest to Camp-II they had to cross numerous snow galleys- climbing some of them and traversing others. Camp-II was established below a rock face over a small overhanging Glacier. The space was just enough for two small tents and they had to do their morning chores with belay on. From there a steep gulley took them to the crest of the ridge that was above the Col mentioned by the Germans.

From the point where they hit the ridge they followed Zemu side of the crest. But there they had to walk on the crest itself with sheer drops on both sides. At that time, the route- opener had to crawl hugging the ridge. After four days work Prem Chand's rope was relieved by Kiran's rope and after two days, on the 30th of April the temporarily established Camp-III at the height of 6,300 meters enabled them to be near the site of work.

From Camp-III to Camp-IV was the most difficult part of the ridge. Normally, a 300 meter climb would take two to three hours. That ridge consisted of towers and mushrooms galore; together with continuous walk over an exposed thin ridge that needed ongoing ice cutting and rope fixing.

Kiran's party consisting of Norbu and Kura Ram worked for two days on the ridge fixing the rope all the way. At one stage, they had to squeeze into a four meter tunnel as there was no other way to cross that section. That tunnel, with a diameter of less than a meter, had to be crawled through and the rucksack pulled with a rope later. Major Pushkar Chand, Havildar Gurcharan Singh and Havildar Nirmal Singh relieved Kiran's party. For the next two days, the weather remained extremely bad and little work could be done. Pushkar Chand returned to Camp-III because of "altitude sickness". Due to bad weather, that rope could not make such progress and Major Prem Chand and Naik Sherpa were inducted again. For the first two days, Gurcharan Singh and Nirmal Singh, who later returned to Base Camp, supported them.

The last ice-tower on the ridge that was one of the big obstacles had not been crossed till the 10th of May. Time was running out. After that ice tower, as seen from Camp I, there were two other towers that seemed more difficult. Their only hope was that there might be an ice field behind the first tower allowing them to circumvent the other two towers.

One day, while watching through a gap between the big tower and the adjoining ice needle, Kumar saw sunlit snow that created hope that there could be such a field just across the first ice tower. That proved right and a difficult portion came to an end after the last tower was negotiated. After Major Prem Chand had opened the route across the ridge, he came down and was relieved by Kiran, whose party consisting of Kushal Singh and Kura Ram who established Camp-IV at the height of 6,630 meters on the 12th of May. The ridge had become a nightmare for the Sherpas. Some of them protested that it was too dangerous and refused to go. Once they took a Sherpa to Camp-IV, he would refuse to come down but was willing to work between Camp-IV and Camp-VI only. One Sherpa remarked that even the southwest face of the Everest was an elephant walk compared to that monkey crawl.

After Camp-IV the going was easy. As one looked at the ridge one could see three humps before the ridge appeared to end and joined the Northern ridge. Though the route to Camp-V was opened on the 15th of May by Kiran's rope, the camp could not be established till the 18th of May, as every time the track was made, it was obliterated by fresh snowfall and wind drift snow. Bull had intended to put that camp somewhere near the last hump but could not do so.

Untimely, Havildar Gurcharan Singh and Havildar Nirmal Singh established camp at a height of 7,230 meters — just below the third hump on the 18th of May. Major Pushkar Chand replaced that rope and he was given the job of pushing Camp-VI up to the Col, between the North- East ridge and the North ridge. However, after the third hump, he was halted by a long thin ridge that ended in a small rock cum ice peak forming the eastern edge of the North East spur. It was the 24th of May and the monsoon was getting closer. By then the problem of high altitude and cold began to plague them. In all, five people suffered from the "high altitude sickness" and two from the frostbite. Even those who were fit felt the increasing altitude.

Once the camps were opened, ferry after ferry went up and down creating a dump there. Sherpas and members came to pick up the stores left by the porters. One day a huge ice avalanche came and buried the entire dump. Luckily, no one was there when the avalanche struck.

Immediately after Camp-IV was opened, Bull started thinking about the selection of the first summit party. For the leader it was a very unpleasant decision to make. If he could, he would have sent all the members up together, but logistical problems did not allow such an easy solution. It was at such crucial decision making junctures that the leader detached his emotional self from his team mates; some of whom were bound to feel slighted. While in empathy with them, he would take consolation in the fact that after all, the endeavour was a team effort.

There are many factors to be considered in the selection of a summit team. Firstly, the summiteers should be in prime physical condition. There are many climbers who, for the sake of the expedition, work themselves to exhaustion and become physically unfit at the time of the summit climb. Secondly, the leader must be fair to the climbers who have worked harder and taken more risks during the climb than the others. The third major factor to be considered is that both the summiteers and the leader should have faith in each other's capabilities. That confidence would not only boost their morale at crucial moments but also help prevent accidents. Fourthly, the summiteers should be temperamentally attuned to each other. Due to scarce oxygen supply to the brain at high altitudes, climbers become irritable. Once, Smyth, the famous British mountaineer, said, "I felt like killing my companion, the only reason for my not doing so was how I would face the public after the return." The last but not the least consideration in Bull's mind was that he wanted the summit party to be a combination of NCOs and Officers who were in equal number in the team.

Major Prem Chand and Naik Nima Dorji Sherpa, to his mind, fulfilled most of these considerations and were selected for the first summit party. He had no doubt as far as Major Prem Chand was concerned, but Nima Dorji had an attack of "high altitude piles" and lost a lot of blood. In fact, once doctors had almost evacuated him out of the Base Camp but Major Sen, their senior doctor, assured Bull that he had fully recovered. Even though there was the possibility of the piles erupting again, Bull decided to take the chance, but made sure that a reserve was at hand to replace him in case there was any emergency that made it necessary to drop Nima.

Once the first summit party was announced, Bull took them off the mountain and sent them down to the Base Camp where Major Sen almost "overhauled" them by giving them all types of food to offset the deficiencies caused by long exposure to high altitude and exhaustion.

Captain Kiran Kumar and CHM. Norbu were selected to support that party and also make a second attempt on the mountain. As the support party they had a hard task and a lot was expected from them. Both of them were taken off the mountain as soon as the route to Camp-V was opened. The itinerary of these two teams had to be carefully worked out to ensure that they did not spend too much time at high altitude before the summit attempts. While they were resting, the stocking of camps was carried on under extremely bad weather. The Sherpas were carrying less and less as they went higher and higher. There were doubts all round if they would ever make to summit. Monsoon showers were forecast for the 1st of June and time was running out.

Bull had intended to establish Camp-VI at the Col but it was placed a km short of that. If it had been placed at the planned height, the last camp would have been set up at 8,060 meters, leaving the summiteers to do only 455 meters on the last day. However, what was done was done, and they had to take their chances with the elusive summit. On the 25th of May the summit party went to Camp-V. At that stage he was forced to induct Nima Dorji to open the route to the north ridge that should have been done by Major Pushkar Chand. In case Nima was too exhausted, Norbu was earmarked to replace him.

The high altitude ferries of seven Sherpas from Nepal, superlative in every way, were sent to Camp-VI to carry loads to Camp-VII. The monsoon clouds were rising. There was a thick, grey carpet spread all over the valley but, from Camp-III onwards, the weather gave them a break in the early hours of the morning. With the race against time and a matter of happenstance, the expedition suffered a nasty blow. There was a 500 metre thin ice ridge between Camp-VI and the Col. Sherpas who were at Camp-VI arrived at the conclusion that it would take at least ten days to fix a rope along the ridge, without which they were not willing to carry loads. Having made up their minds, they came down from Camp-VI.

Kumar vividly remembers the expression on their faces when they told him at Camp-V, "Sorry, Bara Sahib, we must go down." Their moist eyes frightened him – what they were actually saying was that the expedition had come to an end. They had taken tremendous risk; stood by him all along, done a wonderful job and then the time had come to part ways! They came one by one to him, held his hand and looked sadly into his eyes, as if saying, "It is all finished. Call it off". They put on their rucksacks and left. That was the saddest day for Bull as for the first time he harboured doubts about the success of the expedition even if they did manage to clear the route then. They did not have the manpower to carry the loads to Camp-VII and ~~all~~ his plans were shattered.

The next day, the 26th of May, Major Prem Chand and he went up to Camp-VI and beyond to see themselves the obstacle that had sent the Sherpas back. When Bull reached Camp-VI, he vomited. The streaks of blood in his vomit did not alarm him. That was expected at high altitudes. Despite his physical condition, he had to go higher to see the thin ridge. He rested for a while; he borrowed an oxygen set from Prem Chand and went up.

While standing on the wind blown ridge, he felt satisfied that it was not much of an obstacle compared to the route they had already opened between Camp-I and Camp-IV. He could have stood there for hours. He imagined the feelings of the earlier German climbers who had also stood there -in fact, 20 meters higher, and had to call it off. Would the history repeat itself? At last, he returned as he had to go down to Camp-V. At Camp-VI, he had discussions with Major Prem Chand, who was confident that he would hit the North ridge in two days' time. But what about the Sherpas who had to carry the equipment to Camp-VII? With all sorts of doubts Bull returned to Camp-V, where he was to stay for eight days.

The next day, Major Prem Chand crossed the ridge and hit the Col an open place where Bull had intended to put Camp-VI. There he faced the steep face of the north ridge. It was that face that had sent the Germans back. Bull knew he would do that but what about the Sherpas for Camp-VII? They had by then reached the Base Camp and to get those seven best Sherpas back would mean a delay of another four days that they could not afford as the monsoon clouds were ready to burst. Major Prem Chand and Naik Nima Dorji Sherpa crossed the dangerous face on the 28th of May and left five oxygen bottles on the "ring contour"- a small hump on the north ridge. Despite the big setback they had suffered, that was a day for jubilation! They had overcome the last obstacle. There was nothing that could hold them back then except the monsoon and the lack of ferry support. On the 29th of May Prem Chand and Nima rested.

Bull's constant advice to Major Prem Chand was to set up Camp-VII as high as possible. In the existing circumstances it was not possible to stick to their plan to have it at 8,060 meters. However, he decided to have a try. In case they did not make it, the second summit party would shift the camp up. By then their stalwart Sherpas from the Base Camp would also have come up.

Camp-VII was established at 7,995 meters, approximately 575 meters below the summit. Bull had his misgivings but when he compared it to the Everest Expedition, where they climbed 600 meters from 7,880 meters to 8,480 meters – in five hours, he thought they still had a chance. Moreover, Major Prem Chand and Naik Nima Dorji were in excellent physical and mental conditions. Their hopes soared again. He could not speak to Prem Chand directly as they had small wireless sets, but both of them could speak to Major SS Singh at Camp-I as he had a bigger set. Major Singh was controlling the signal net and also looking after the ferries up to the Camp-III. There Captain Cruz their second doctor was managing the loads, at Camp-IV young Captain Bahuguna was looking after the logistics; and at Camp-V, Bull had established his headquarters. At Camp- VI, Captain KI Kumar was in support. It was a pity that Bahuguna, who had done excellent work during the last stage of the expedition, got frostbite and had to be sent back. Cruz took his place and Major Sen moved up to Camp-III. Their numerical strength was so short that doctors had to be given the work of climbers.

On the 30th of May Prem Chand and Nima Dorji spent a very uncomfortable night. Their tent was torn by the wind. Their stoves would not burn owing to the gale. However, they were able to melt some snow to make tea and slept as they were. The morning was hardly better on the windswept ridge. When they tried to put on their oxygen set they found one of the regulators leaking. Luckily, they had carried a spare regulator that was usually not done; which saved the day.

At about 5:00 am on the 31st of May they left camp and started climbing. They crossed the snow field in which their feet sank 30 cm. They thought of going up the west ridge but later decided to stick to the north rocky ridge. The strong winds had also left their mark on the stones that were bare of snow, and showed heavy erosion.

The Sherpas who had left the summit party at Camp-VII had told them that it would not take the climbers more than for four to five hours to get to the summit and so, when they did not get any news until 4 p.m., Bull Kumar started worrying as the climbers had left their wireless set at Camp-VII. As Major Prem Chand narrated later, at 12.30 p.m. came the time to make a decision; the summit was not in sight; should they carry on, or return? Carrying on meant danger! If they could not get back to Camp-VII before nightfall, they would have to spend the night out on the windy ridge! If they returned, three months of hard work, the tremendous risks taken, a life lost would come to nought! They took the decision to carry on.

As they climbed further they suddenly found themselves on the summit ridge of the Kanchenjunga dome. At 2:45 pm they got to the summit. In reverence to the Sikkimese sentiments – they worshipped that mountain – the summit party left the last two meters to the top untrodden. They fixed an aluminum piquet near the summit and hoisted the Tri colour and the Army flag. Then they took out other flags of the Indian Mountaineering Foundation, and the regiments they belonged to and took pictures.

After some time, they started back. The weather all around except on the ridge, was bad; at one place they had a fall and rolled down for several meters. Then the night overtook them. The sun sets early in the east. Once again their luck held – a ray of moonlight lit up their ridge. Finding their way with difficulty, they reached their camp at 8.15 p.m. It was a great relief not merely to them but also to the rest. Their descent from Camp-VII was not an easy one as Nima Dorji perhaps suffered from snow blindness. He could see with great difficulty and pain. On the 1st of June they reached Camp-VI where they were welcomed by Kiran and the others.

The second summit party was poised at Camp-VI and was to move out on the 2nd of June but when Bull awoke he found the cloud base had risen from 6,960 meters to 7,574 meters. He did not want to taint success with tragedy and called off the expedition. It was fortunate that he did so, as the very next day, they had heavy snow fall and could only come off the mountain with great difficulty. On the 6th of June they reached the Base Camp abandoning a lot of equipment on the mountain. When they reached Lachen, the road head, there were heavy rains and the road to Gangtok had been washed away. They were waiting for road to open when they got a message that the Vice Chief of the Army Staff, Lieutenant General OP Malhotra, was coming to receive them on the 16th of June. It was a matter of great honour for Bull and his team. They took their boots out once again and started marching, leaving behind their entire luggage. After spending some time at Gangtok, they reached Delhi on the 24th of June.

The summit party had taken two cameras and two films with them. When they came down from Camp-V Bull handed over one film to Major Prem Chand as he was responsible to prove to the world that he had climbed the summit of Kanchenjunga. Bull thoughtfully mixed it with all the other films so that nobody could find the summit film.

When they reached Darjeeling, they got busy with parties and celebrations. In the meantime, they got a message that the Chief of Army Staff wanted the summit picture immediately. So, he asked Prem Chand to give the film to Das Studio for development. He was shocked to hear that somehow that film was totally blank. If they did not have a photograph from the summit, it was difficult to prove their claim, especially on a route that had defied mountaineers for 46 years. Bull was really worried and he prayed that the second film that he had mixed with the other films should show some results.

He did not inform anybody in Delhi about the first film as he did not want Army Headquarters to get worried. The Chief of The Army Staff had put all his eggs in his basket against the advice of the Chairman of the Indian Mountaineering Foundation, and he would have sleepless nights if he came to know that the summit pictures were not there. When he came to Delhi, Bull gave all the films to Mahatta's for processing. Madan Mehta of the Mahatta Studio informed him that the technician who does the colour film was on leave and the processed film would be available after only a week. He informed the Vice Chief about the same.

There were many receptions for their expedition and in some places they were also handed over cash prizes from many dignitaries, but he did not enjoy any of these receptions as he was worried about the results of the second film. Exactly after 7 days, General OP Malhotra, the Vice Chief asked him for the summit photographs. That was the day when Mahatta Studio had promised to process his film. He quickly got ready and went straight to Mahatta Studios in trepidation about the fate of the film. Lo and behold, guess what Madan Mehta told him, "Sir the first film we developed was that of the summit and you have got excellent shots." Bull let out his sucked in breath in relief!

The Chief of Army Staff, Gen. TN Raina, MVC had already taken an appointment with the Defence Minister to show him the summit films. Shri Jagjeevan Ram, the then the Defence Minister was exuberant when he presented him a large frame picture of the south summit of Kanchenjunga - irrefutable evidence that they had climbed the summit.

TERAM KANGRI– 1978

After 1977 Kanchenjunga conquest Bull Kumar, "Bull", felt that a great vacuum had crept into him. It would have been unbearable to let a climbing season go without being in the Himalayas and yet nothing looked challenging enough after the North East Spur of Kanchenjunga. All of a sudden he had another goal and once again he was alive with adrenaline flowing into his blood stream.

In 1978, his German friend Volker Stalbhom from the Indus Boat Expedition came back with an exciting proposal - to lead another expedition, this time down the Nubra River. Bull Kumar They showed him a map and told him that though that area was in Pakistan, no Pakistani had ever come down to the snout of the Siachin glacier. Taking one look at the map, Bull knew that there was a cartographic error. He purchased the map from the Germans and went straight to Director General Military Operations (DGMO) Lt. Gen. ML Chibber. He knew him personally as he had served under him. Bull explained to him about the cartographic error but he was too busy to even look at the map and he sent for his Assistant Director General Military Operations (ADGMO) Brigadier Mehta to listen to his problem and also told him to give Bull a cup of tea as he had been an officer serving under him.

Brigadier Mehta and Bull went over all the treaties starting from the Sikh-British War, till independence, right up to Simla Agreement and in the next two hours he requested the DGMO Lieutenant Genereal ML Chibber if he could see them. When they met him, all other work in the Military Operations Directorate stopped. The cartographical error had ~~got~~ caught everyone's attention then. The DGMO asked Bull how he could help them. He proposed an expedition to the Siachen glacier and as he was the Commandant of The High Altitude Warfare School, he decided to lead the advanced mountaineering course for that expedition.

Bull read all the literature he could lay his hands on, and found that Siachen was the longest non-polar glacier in the world. Looking around he found that that area was also the most glaciated outside the poles. Someone had given that the name of "third pole". Around these glaciers were hundreds of peaks. He selected Teram Kangri-II (24,300 feet). He became a live wire again. That was by no means going to be as challenging a climb as the Kanchenjunga, but he knew that it would be a different experience -- an experience of an exploration!

He did not know how other climbers select their objectives. With him unless he got an inner call, the acceptance of an idea, it meant nothing. But studying Siachen glacier breathed new life into his earlier listlessness. The post Kanchenjunga time was rather busy for him. The expedition had to be wound up, the book had to be written and above all he had an all India lecture tour besides having just taken over as the Commandant of The High Altitude Warfare School.

There was just not enough time to do all that and simultaneously prepare a venture of the kind he had in mind. His proposed expedition entailed crossing the entire Nubra Valley, Siachen Glacier - the longest glacier in the world and then get to the crest of Karakoram where Teram Kangri stood. With the time at his disposal it was an ambitious project. He needed administrative support on the same scale as on Kanchenjunga if not larger in magnitude.

So he called on the new Chief of the Army Staff (COAS) General OP Malhotra, PVSM and unfolded his plans. He was relieved to hear "Karakoram, excellent idea!" He had nothing to worry about. The COAS stood like a rock behind him and sky was the limit as far as backing was concerned. In the time available, he rushed around collecting equipment, rations and information. Major Awtar Singh and Major Mehta of The High Altitude Warfare School were the back bone during the planning stage so were Major Stobdan and Captain Mann during the conduct of the expedition.

The main party reached Khalsar in the Nubra Valley on the 2nd of September 1978 after crossing the highest motorable pass of the world Khardungla (18,300 feet) on the Ladakh range. The journey from Khalsar to Base Camp was not an easy one. The biggest obstacles encountered were the unprecedented floods that had affected Ladakh in 1978. Neither in living memory nor in history had Ladakh ever experienced such massive floods. Many times their horses and camels, that carried load for them, got washed away in flooded water of the Nubra (Yarma) river. In all, their loads were carried by 300 horses and at places they had to camp by the river side for water to subside before fording. Luckily, he could get the horses across. The Nubra Valley was much greener than Leh Valley and at places most unexpectedly they did find thick forest of thin bushes.

On the first day they passed through Tegar village. The highlight of their visit here was lunch given by Major Stobdan's (Deputy Leader) father. In all his life, Bull had never drunk cleaner and tastier 'chang' (local beer). From here they employed five double humped camels. The ancestors of these camels had come along with the Central Asian traders over Karakoram pass. At that time there were only 17 of them surviving in the whole of India and all in the Nubra Valley! About six miles ahead of Tegar, they encountered Solan Chamesshan Lungpa that came down from Saser Kangri (the Golden Summit). Just across that they could see the old palace of the former ruler of Nubra at a place called Chalsa.

They had an enjoyable well-deserved bath at the hot springs of Panamik. Panamik was a huge village where the Central Asian traders had halted to recoup and recover after a 30 day journey from Yarkand over the most inhospitable highland of display and the high pass of Karakoram. Ten miles ahead of Panamik was Sasoma where the old silk route from Central Asia met Nubra valley after descent from Saser-La. They were told that the trail over that route was marked by the skeletons of camels and horses that had fallen prey to high altitude, severe cold and exhaustion of the treacherously harsh journey.

From Sasoma to Tongstead the track passed over scorched and came to rocks that had been termed as "sheer rocks" on the map. Nubra River had about six branches and the Eastern most branch was waist deep and 50 feet wide cutting along the steep rock. They had to erect a "flying fox" to get across their loads across. The horses and the camels had to swim across. The next two km trek went through water-logged area and every one had to cross numerous streams. It was better to hold their boots in their hands. Tongstead village was the last main village before one reached the snout of Siachen Glacier. It had its own outpost at Warshi, that was four km away and had a population of 3 persons; husband, wife and daughter. They called on the Mayor of Warshi where they were given lovely salted butter tea. They established their Base Camp at Pra about half a km short of the Siachen Glacier.

Most of the snout was passive and looked like a tongue. Between the Base Camp and the tongue of the Siachen Glacier there was the fast flowing Nubra River from the icy caves. To negotiate that obstacle the team tried to use reconnaissance boats but that was not considered safe since the current was too rapid. The "flying foxes" were also tried. These were successful but time consuming. The only way out was to climb the rocks east of Siachen Glacier. The route was made safe by fixing 1000 meters of handrail rope.

Stage-I was put on the glacier in the area where the U turn of the Glacier ended. The leader was successful in carrying out a reconnaissance to go straight from the snout which was shorter by 2 km. However, that route proved dangerous for the porters due to numerous crevasses. It was decided to stick to one on the medial moraine and move along the grain of the Siachen Glacier. Apart from three or four bad crevasses that were marked, the route became fairly safe for the porters. Route to Stage-II was opened on the 9th of September 1978. For the first 3 km or so the track ran along the grain and was safe which other places was obstructed by glacial streams. At places it came across non-negotiable crevasses and had to be crossed over to the western side of Siachen Glacier. Many porters and members slipped and cut their hands on sharp ice. Stage-II was established three miles short of the junction of the Chumik Group Glacier and the Siachen Glacier.

For administrative convenience, Stage-II was converted into a dumping point. Loads carried from Stage-I were dumped at Stage-II. These were picked up by the porters from the Stage-III. That system had great advantage as firstly that reduced the number of stages that they had to maintain and secondly ensured flexibility in directing porters from one stage to another.

The entire route between Stage-II and Stage-III was strewn with huge boulders and was extremely rough. Stage-III was placed after the junction of Siachen Glacier and the glacier coming from the Chumikchan group. The glacier was covered with small stones and there were no crevasses. Stage-IV was established in an area one km short of junction of Siachen and Lolofond glaciers. From Stage-IV it was planned to cut across Siachen Glacier again and have their camp in the Lake area. However, two deep streams bordered that approach almost 50 feet wide; and despite all their efforts they could not cross them. The advance party went higher and luckily found a huge snow bridge over the streams and hit Teramsher Glacier almost about three km west of the lake and established Advance Base Camp. That was one of the roughest routes encountered by the expedition; they also had to negotiate a small ice fall there.

After crossing a number of different coloured moraines they climbed about 500 feet of hill on the left bank of the Teramsher glacier where Advance Base Camp was established due south of Teram Kangri. In retrospect the new Advance Base Camp that was about an hour's journey from the planned 'Base Camp' was more practical than the Lake Base Camp. No doubt the Lake Base Camp had advantages of grass, water and shelter. The Advance Base Camp was five days march from the Base Camp and to get supplies for seventy people with only 50 porters was a difficult task. They were almost living from hand to mouth, but work went on.

Camp-I was established in an area where the glacier coming down from Teram Kangri-II starts at the height of 19,000 feet, after crossing the Teramsher Glacier that was about 3 km wide and full of hidden crevasses and pot holes. In the early stages when the weather was warm, the biggest hurdle was in the form of puddles of water, rendered dangerous as they froze due to low temperatures at night making a thin film camouflaging the crevasses. It was fortunate that nobody broke any bones as these were real traps!

From Camp-I that was just below the south face of the peak there were routes to the summit, one from the south West ridge, second diagonally across the face from east to west and the third straight up from the right. The entire face was fastened with dangerously hanging glaciers that came down destroying everything that came in their way. Having watched the trend of avalanches for seven days Bull came to the conclusion that the right rib route was the safest of all.

Route from Camp-I to Camp-II was over a steep rocky ridge. Major Stabdan and Havildar Puran Chand's parties opened the route in turn. They established Camp-II at the height of 21,000 feet. The entire route was roped up. Part of it went under a loosely hung ice mass that posed a great danger. That was the calculated risk that the climbers had to take. There was no dearth of avalanches but they were lucky as the avalanches did not strike when the parties were working.

Camp-II was the most uncomfortable camp of the entire expedition. They had to make retaining walls of stones to make platform for the tents. The worst part of the stay in Camp-II was toilet; in the morning that had to be done holding the rope and balancing oneself very precariously and at the same time watching for the falling stones from the above. One of the Jawans was hit by falling stones and came down unconsciously. Luckily his friend who was just below him held him. The average climbing time from Camp-I to Camp-II was four hours.

From Camp-II onwards there was no camping place that could be considered safe. After 500 feet of climb one had to cross a small ice fall where a huge crevasse 50 feet wide and 100 feet long lay across the route that was difficult to negotiate even with the fixed ropes. After that the route lay on the small rib that ended on the rocky face just below the great ice fall. The team established their Camp-III in a stony area at the height of 23,000 feet.

The route to Camp-III was opened on the 6th of October 1978 and the first summit party along with the leader went up to Camp-II on the 7th of October 1978. It was decided that the party supported by six would go up to Camp-III and make the attempt on the peak. The next day the weather turned nasty and the party had to return to Camp-I. During the night they heard an explosion and their tents shook as if there had been an earthquake. It was an avalanche; luckily the avalanche passed 50 feet below the camp and they were safe. For the next few days the weather remained hostile and no work could be done on the mountain.

The second summit party consisting of Major T Stabdan, Captain MP Sharma, Havildar Angraj Singh Pathnia and Dhujman Rana supported by four other members again occupied Camp-II. Captain MP Sharma's foot hold gave away and he fell 50 feet down the rock. He had a miraculous escape from certain death. He was evacuated by helicopter and Captain KK Sharma took his place.

On the 7th of October 1978, the party reached Camp-III. The next day was supposed to be the summit day but the winds were too strong for any movement; and temperatures extremely low. The party could not climb even 200 feet. Havildar Dhujman Rana, who was the strongest from the The High Altitude Warfare School and could carry 60 kgs of weight at an altitude of 19,000 feet, felt that all his energy had drained away by the biting cold. It was very wise step on the part of Major Stasbdan to call off the attempt.

It was a difficult decision for the leader to make at that juncture whether to make the third attempt or to call off the expedition. The Leader called the party to Camp-I and personally inspected the clothing and physical condition of all the members before taking the vital decision. Two cases of frostbite were noticed and were evacuated.

Due to extreme cold and strong winds it was then decided that the minimum number of people should be exposed to high altitude. Therefore, the last summit party was not given any support party up to Camp-III. However, a support party was kept in readiness at Camp-II for emergencies. The last party consisted of Havildar Puran singh (Para), Havildar Awtar Singh (J & K Rifles) and Naik Kalam Singh (Garhwal Rifles). That party reached Camp-III at 2.00 pm on the 11th of October. It did not waste any time and decided to fix 200 feet of rope to make things easier for the next day -- the summit day. That decision of Havildar Puran Chand had far reaching consequences on the final attempt.

On the 12th of October 1978, the party though ready to start at 5.00 am could not do so due to extreme cold and had to wait for the sun to come out. They left the camp at 7:00 am and they made excellent progress. However, then they came across a cornice that was protruding 50 feet high out of the hill. Havildar Puran Chand found a small hole; entering it he found a big crevasse. He climbed through the roof of the crevasse to get to the top of the cornice - an excellent feat of mountaineering! The time was 4:00 pm, the ice wall had wasted four hours and the peak was 2 km away. The party returned to Camp-III. Every inch of their movement could be seen and the others' hearts sank when they saw the party coming down. The entire night passed in anxiety.. According to the plan the party was supposed to come down the next day as only two days' food was carried.

But on the morning of the 13th, they saw the party going up again. It was a great relief. At 11.00 am the party was on the top of the crest of Great Karakoram Range. Now Bull knew they would succeed. The summit party found a beautiful plateau on the other side of the crest. In four hours they covered about two km of easy gradient to the summit. They were seen on the summit from the Advance Base Camp and Camp-I. Where they stood, with the turn of their heads they could see Afghanistan, Russia, China, Pakistan and of course India. None of these countries were more than a 100 miles away. That could be the great vantage point of Asia.

So, on the 13th of October 1978 at 4:45 pm, three soldier mountaineers stood atop 24,300 feet Teram Kangri-II Peak on the main crest of the great Karakoram Range. The snow under their feet had different destinies. On one side, it formed into a stream that would become the major tributary of the Yarkand River that after wandering through China's Kun Lun Mountains and the fabled city of Yarkand got lost in bowels of Gobi desert. Yet the snow on their side of that greatest watershed of the world would merge into the Indian Ocean via the Nubra, Shyok and the Indus rivers.

It was a politically important event. The soldier mountaineers were standing in the Shaksgam valley - one of the most inaccessible regions of the world; that not more than a handful of people had dared to enter. That was the region that Pakistan had ceded to China in 1963, even though the area belonged to India.

It was also important for the territorial integrity of India, as it was the first time the Indians crossed the "Lakshman Rekha" drawn by the Americans in their maps connecting Point NJ 9842 to Karakoram Pass. Just with one stroke of a 5 inches long line, they transferred over 5000 sq km of Indian Territory to Pakistan. It was also important because Indians became aware that the Siachen Glacier "the longest Glacier in the world" did exist and belonged to India. It was India's first face climb. Normally Peaks are climbed from ridges. Face climbs are more difficult and require technical know-how.

The Shaksgam branch of the Yarkand has its sources in the Urdko, Kyagar and Kashebrum glaciers of the Karakoram (records geological survey of India LXII, p.263), and it drains the Central Asian watershed at E, Chart XXXV. It was surveyed by Major Mason in 1926. (Exploration Shaksgam valley: records, survey of India, XXII). The Shaksgam River, (called also Oprang) follows a long course north of Karakoram parallel to the course of the Indus in Ladakh: under the name of Zarafshan, it passes through the Kunlun Mountains at their North-Western termination, against the meridional range of Muztagh Ata. It flows across the Takla Makan Desert past the city of Yarkand, and after it has been joined by its Kashgar tributary it is known as the Tarim River. Like the river Chitral the Yarkand River flows under many different local names that are at times confusing.

After taking pictures of surrounding areas and hoisting the flags of the Indian Army, The High Altitude Warfare School and their regimental flags the summiteers came down reaching Camp-III at 8.00 pm -- indeed a stupendous achievement and Havildar Puran Chand had to be commended for all efforts.

Col. Kumar presenting the Ice Axe which he planted at Indira Col to Indira Gandhi

Sia Kangri – Tri junction of Pakistan occupied Kashmir, China occupied Kashmir and India

Col. Kumar at Sia La

Ice avalanche on the way to Saltoro Kangri

Teram Kangri – 1978

Col. Kumar on the Summit of Sia Kangri

Col Kumar with the summit party of Saltoro Kangri

Col Kumar at Indira Col, the source of Siachen Glacier. On the back is the Shaksgam Valley which Pakistan has ceded to China

SIACHEN – 1981

Once Bull returned from The Teram Kangri – 1978 expedition he suggested to the Army Chief that the cease fire line should be drawn along the Saltora Range from the Point NJ 9842 right up to Sia Kangri. He also suggested a second expedition to Sia Kangri, Indira Col and Saltora Kangri – the highest mountain in the eastern Karakoram should be undertaken. He got no reply for two years. Dejected, Bull decided to resign. His resignation was accepted. He had six months of leave before his retirement from the Indian Army. In those days, leave encashment was not an option.

Just as he was to begin that leave pending retirement, his proposal was accepted and HAWS was asked to prepare for the expedition. Since that was his proposal to begin with, Bull wanted to see that through to the very end. As a result the Siachen mission was carried out during his 6-month leave pending retirement. Since he was in medical Category 'C' permanently - that meant he could not be posted above 7,000 feet due to frost bite in his previous adventures, every mission he took was at his own risk – and he had to sign a bond saying so.

In April 1981, he went back to Siachen Glacier. As that expedition had the blessings of the Army Chief, raising of funds and equipment posed no problem at all. Most of the equipment used was indigenous. However, some foreign equipment was purchased from Nepal. The team collected in Gulmarg in the first week of May and went through strenuous conditioning ranging from 5,000 to 9,000 feet of climbing every day.

Since the Khardung La (Pass) 18,300 feet, the highest motorable pass of the world (on Ladakh range) was still closed due to heavy snows, the Advance Party flew into the Nubra Valley from Srinagar. Bull along with the main party, got to Leh via traditional Kargil road on the 8 June. The main party had already crossed three mountain ranges – the Pir Panjal, the Great Himalayan Range and the Zanskar Range and had come 400 km from Srinagar. Their destinations lay another 300 km north of Leh. To get there they had to cross two more ranges - the Ladakh and the Karakoram Ranges.

As they approached Leh, before them lay a glorious view of the Indus River streaking out in silver lines. Leh lies at 11,500 feet at about the same height as Lhasa. It is an arid, brown and windswept town at the cross roads of the ancient trade routes with Yarkand to North, Lhasa to the East, Punjab to the South and Baltistan to the West. The most difficult of all the trade routes, while also the most important, was the silk route from Yarkand. After crossing the 18,300 feet high Karakoram Pass it went over Saser La 17,480 feet (Karakoram Ranges) and then Khardung La 18,300 feet before reaching Leh. Along that route traders from Chinese Turkistan used to bring silk, carpets, corals, jade and took back with them cotton, brocades, opium, indigo, turbans, tobacco and also barley, rice and wheat.

As one enters Leh, the first building to catch the eye is the old royal palace – Lehkhar, seven stories high. In the shadow of the palace in the market lies the mosque built in the seventeenth century by the orders of Aurangzeb to mark an alliance between the Ladakhis and the Moghuls against the Tartars. The Raja also embraced Islam and became Akbar Mohammad. Another prominent monument is the Leh Fort built by General Zorawar Singh – the Dogra Conqueror of Ladakh. Leh is full of chortens, dedicatory buildings, erected in the honour of the holy ones. The huge chorten at Leh near the Bazar is perhaps the most photographed monument of Leh. Just across the river Indus lays the Stok Palace the only palace in Ladakh that is still inhabited.

Most members of their team had come to Leh for the first time and went on a whirlwind trip to see the Gompas – Shey, Thikse and

Hemis. Hemis is the largest and wealthiest monastery of Ladakh. It was built only a century after the Potala Palace of Lhasa. It is said that the riches of Hemis are next only to those of the monastery of Potala.

After a couple of days' stay in Leh they moved over to the Nubra Valley after crossing over the 18,300 feet high Khardung La (Pass). Border roads have done a tremendous job of clearing the 40 feet deep snow that made the pass inaccessible for the greater part of the year. Nubra Valley lies along the Nubra and Shyok rivers and is approximately four rocky faces, rising up seven to eight thousand feet and culminating in snowy peaks of 24,000 feet and above.

They followed the valley that was along the Nubra River and was flanked in the West by the Saltoro range, in the east by the main crest of the Karakoram Range, in the South by Ladakh range, in the North it ended in the Siachen glacier that is the longest glacier in the Himalayas and Karakorams put together. At the head of the Siachen glacier are the Turkistan La and the Indira Col that form the water shed between Central Asia and the Indian Sub Continent.

They crossed the Shyok River at Tirit and went along the great silk route passing the villages of Semur, Tegur and Panamik – well known for its hot springs. Panamik is the last village on the silk caravan route till the trader reached Eastern Turkestan. The caravans had to take 20 days' supplies with them before leaving Panamik to enable them to cross. At Sasoma the silk route branches off to go over the Saser La Pass but they carried on towards North to Tongstead village.

It was there that they found the double humped Kirzig camels. These were the replacements of the yaks of the Eastern Himalayas. Their next halt was again at Warshi – the last village of the Nubra valley that consisted of only one family – a husband, a wife and a daughter.

Throughout the day's march the eye wearied with the dull monotony of sand, stone and more sand through which they had been passing. It then rested with delight on the terraced fields of green grain bordered by a profusion of poplar trees and rose bushes. Warshi is situated in the centre of that oasis in the wilderness. It is a small village. These gardens in the desert are created on every bit of alluvial soil that has been retained by stone walls in terrace formation.

A mountain torrent nearby furnished a never failing supply of water. The size of these villages normally depended on the industry of the inhabitants. Considering that there was only one family in Warshi they must have had really slogged to make that small heaven.

On their fourth day of marching they reached the ice cave of the Siachen Glacier where the glacier ceases and gives birth to the Nubra River. Here they found abundance of wild rose bushes that had given their name to the glacier, Sia in Ladakhi meaning "Rose". These wild bushes often grow luxuriantly amongst boulders where no soil could be seen, or high up on the faces of the perpendicular rock precipices, the colourless surface of that they relieved in the most fascinating manner. They looked like gems in the sandy and stony wilderness. Each tree or group brilliant with every shade of mauve – from the palest pearls to the deepest red; but were surprised by the gushing waters of the river as well as ice chunks of glacier that had fallen in their attempts. They climbed up the steep rock faces to get to the tongue of the glacier.

The first day's journey followed the rocky moraine troughs that were lined by icy ridges on both sides. That day's halt was immediately after the U-turn of the glacier. The most striking sight here was the disappearance of the huge stream of Terong Tukpo beneath the Siachen glacier. Undoubtedly, once upon a time the Terong glacier had extended right up to its junction with the Siachen, but later it had receded at least 10 km to the east leaving behind a flat brown valley and the river.

In that placid Siachen Glacier hundreds of other glaciers and streams from the east and west had joined to submerge their identity. It was over 1000 feet deep four km wide and 90 km long. It had its own system of drainage above and sewerage below the surface. At times the blue stream just plunged into walls of ice hundreds of feet deep not to be seen again. There were huge tunnels in that glacier that were once the waterways.

On the second day's march they had to cling on to the steep ice banks of those deep gushing streams. One wrong step would have resulted in certain death. According to experts it would take 3 minutes for anyone to freeze, if fallen into the ice cold waters of the glacial streams. One of the porters in earlier expedition had fallen and was frozen to death. The second stage on the glacier was on the rocky moraine at the junction of the Siachen Glacier and a glacier coming down from a 24,350 feet high peak in the west. That magnificent mountain was referred to as peak 8 in the survey of India maps. Of all the Karakoram peaks perhaps that could be seen the farthest.

Another two days' march took them to the junction of Siachen Glacier with Lolofond Glacier. Lolofond Glacier ends in the Saltoro Pass that is the gateway to Baltistan. All expeditions to that area had used that pass for entry into the Eastern Karakorams. So far they could not put their skis to any use as the glacier was full of moraine debris. But there the terrain smiled on the skiers and movement became faster with skis on. Their next halt was the junction of a huge glacier coming from Saltoro Peak on the west with Siachen Glacier. That became their Advance Base Camp (16,500 feet). It was from there that they launched their ski tours to the various parts of the Siachen Glacier.

The first trip was made on the 24th of June to Bilafond La or the Saltoro Pass – as it was sometimes known. They skied down early in the morning to the northern end of the Lolofond Glacier. As was always the case, at its junction with the Siachen Glacier the area was full of searches and crevasses and for about a hundred yards they had to take off their skis to avoid going over the huge boulders in the moraines. But once the junction was crossed they found the going smoother till they hit a Glacier stream. A glass like bridge that seemed strong enough to take their burden, covered the stream. But appearances could often be deceptive. The moment Bull got to the centre of that bridge, the glassy crust gave way and he found one of his legs dangling in the icy cold waters of the stream below.

He got a terrible feeling as the cold waters trickled slowly into his boot. In those few moments he felt an overpowering sensation of numbness taking hold of his leg. He managed to pull out his frozen leg from the stream and removed his skis, boots, and socks immediately.

His buddy Vinod rubbed his leg vigorously to restore blood circulation. Almost an hour was wasted before they could move again at 10.00 am

The slopes were nice and gradual and the climb almost effortless. Bull was sure that they would be able to get to the top of the pass by 12.00 pm. at the latest. At 2.00 pm their destination still seemed another half an hour away. But distances in the mountains could be most deceptive for it was not before 4:30 pm that they reached the summit of the pass (18,200 feet – 5,520 meters) completely exhausted.

What a reward awaited them! The view from the top of the Bilafond La was breathtaking and glorious! To the west lay the magnificent Saltoro peaks while to the East was the most imposing group of Teram Kangri Mountains. Longstaff first visited the pass in 1909 from the Western side. The Saltoro Pass was originally considered as the main watershed between the Indian sub Continent and Central Asia, till Longstaff disapproved it. His measurement of the height of the Teram Kangri Mountain created a furore in the Survey of India office at Dehra Dun. Longstaff's readings showed that Teram Kangri was higher that Everest. Later on, Teram Kangri turned out to be only 24,340 feet high. Vinod and Bull had been to Teram Kangri-II three years earlier. For them that was a view full of nostalgic memories and sentiments.

The mountains on the other side of the pass, though lacking in grandeur and size, broke the dull monotony of snow and ice, as their eyes rested on the jungle of rocky pinnacles piercing the sky. They were lucky without photography as the weather was good. After they had their fill of the spectacular vista of snow, ice and rocks and having gulped down a cup of tea they turned their skis downwards and kept gliding down effortlessly at 15-20 kmph with their cross country skis whistling gaily all the time. To reduce their speed they just had to open the flaps of their jackets. Without taking a single turn they came down for eight long km. The ecstasy of that delightful ski down trip – was one that Bull Kumar Bull would never forget.

They had to wait for two days till their supplies fetched up and then started off on their expedition to Indira Col. They improvised a sledge by fixing a plywood board on the skis. Six of them then went self-contained for six days hauling their supplies like huskies of Eskimos.

They left their ABC (Advance Base Camp) on the 27th of June and within an hour they came to Junction of Saltoro – Siachen Glacier and got a magnificent view of the Saltoro peaks, while Teram Sher Glacier lay to their right. Their progress was excellent as the gradient hardly rose a foot in 30 feet. But soon obstacles came their way and they had to cross one big stream after another. It took them considerable time to find really solid snow bridges that could stand the weight of their sledge. At times the sledge had to be unloaded and taken across empty for reasons of safety.

Two hours of travel took them to a highly creased area. They tried to circumvent it but could not, due to the parallel streams on both sides. Every few minutes the sledge broke through the thin snow bridges and got stuck in the crevasses. Luckily the crevasses were not more than 2-3 feet wide and the sledge did not plunge in. They had to wrestle with the snow continuously to keep the sledge moving. At times, they even thought of abandoning the sledge in utter disgust.

By 4.00 pm they were out of that mess of potholes, crevasses and soggy snow and headed north towards the rocky dome-like structure below Indira Col. An hour later they halted for the night near an icy pond of water that was immediately east of a hawk like peak. They decided to call that peak 'Lhagri' (Eagle).

The next day they crossed the sea of ice coming from the western source – Sia La. The Eastern border of Siachen tapered down to meet the glacier from the heights of Teram Kangri. The grey granite and gneiss structure gave way to yellow crystalline limestone. That black dome like rock feature – a prominent landmark – seemed to them only an hour's journey away. It turned out to be eight hours of hard skiing and pulling of their sledge behind them to reach its bottom. Due to lack of atmospheric pressure and distances' perception at those altitudes became skewed; and as there were no landmarks to break the expanse of ice and snow the best any climber could do was not to make any guesses. That black rock structure was the place where Bullock Workman's expedition had also camped in 1912.

The next day, on the 29th of June, as per the route given by the Workman expedition, they climbed up a steep ridge. There they looked back and saw a wonderful vista of the Siachen Glacier flowing away into the distant forty miles. The first half portion was a huge field of white and pure snow. Later on, the glacier became ribboned with moraines of different colours as if the entire glacier had been ploughed. From a height the sight appeared to be that of several roads running parallel to each other. Across the ridge they found another snowfield. Thanks to their skis the crevasses of that field could not play merry hell with them as they did ~~to~~ with the American team of 1912.

As they skied up the last ice ridge they could see the distant peaks of Central Asia. And then they slowly gained the summit and hoisted the Tricolour on the heavily corniced ridge - to the north they could see low, arid and brown mountains with red slashes – to the south was the jungle of white snow covered high mountains. Immediately below the Urdok Glacier fell way towards north! That glacier ultimately fed the Yarkand River. Immediately towards the south lay the expanse of Rose Siachen Glacier. The winds were terrible in that gap and at that time were blowing from the south to the north. After having spent almost an hour on the northern most point ever reached by any Indian, they skied down. Within a few minutes they had descended 500 feet and in the valley they felt protected and warm.

While they munched their lunch, Bull felt the oddness of the name given by the Workman's expedition to that ridge-"Indira Col". Col is normally the lowest depression in the ridge. In that case it certainly was not so, as they could see that tapering towards west. Right in front of them, they could see the eastern face of Sia Kangri that had brought down tremendous avalanches of red debris and made the whole hill side look like a bleeding mountain, or like the tears of the Rose Glacier that was about to leave the Indian sub Continent and jump over a precipice into Central Asia. As they had plenty of time at their disposal, he decided to follow the eastern face of Sia Kangri.

Though the avalanches on that mountain face terrified them, they calculated that the basin was big enough to hold both, the avalanches and them. And lo and behold, at 2:30 pm they had discovered the real Col on that water parting. That lay in between the ridge coming down from Teram Kangri and the one coming down from Gasherbrum. Workman's expedition had proved that in 1889 Young Husband's team had never got to the Col as claimed by some British authors. Not in any way discrediting the efforts of the American expedition, he could safely say that the Americans did not get to the real Col, if the terminology was strictly applied to that feature. But that would mean splitting hairs. In that immense spread of thousands of km of ice, 3 km was meaningless. After that they skied directly to their Black Rock Camp circumventing the ridge that they had climbed earlier. They were deeply satisfied with the day's work.

The next day they climbed the high plateau again and turned east. Going over easy slopes, in two hours they were on Turkestan La – 18,800 feet. They were lucky to get a glimpse of the snowy peaks across the pass before it got cloudy. Down below they could clearly see the Staghar Glacier running from the south east to the North West. In two days they had been to two major glaciers of the Shaksgam valley. They soon started the downward journey. The snow was still hard and they had to take small turns to control speed of the skis.

At about 18,000 feet they saw a yellow bird merrily resting on the icy slopes. In the beginning they thought it to be a piece of yellow rock but that suddenly flew off as they got nearer. It had long bare legs like the water birds and looked very frail. The wingspan of the bird was about 24 inches. Bull had yet to find out what it could have been. They broke their Rock Tower Camp and proceeded directly to the next camp. On the 1st of July, Bull returned to their ABC (Advance Base Camp).

After a day's rest at ABC, he skied up again to the upper reaches of Siachen, and got to Sia Kangri. Their Camp-I was on the Siachen Glacier situated at a height of 17,800 feet and was named Andaman. It lay in the shadow of the Sia group of mountains. Then the team turned west towards Sia La and followed the western face of the Sia massif. They camped below Sia Kangri-II that had been named by the Americans as the Queen Mark Peak. They had to cross a heavily crevassed area. Going a little higher towards Sia La and then turning right could have circumvented that. But they found that out only when the crevasses opened up later.

From the camp they had a magnificent view of the twin 24,000 feet high peaks and the silver plateau to their west. These peaks were named Mount Ghent by Workman's expedition, but that name had not been accepted by the Survey of India. Other names not accepted by the Survey of India were King George V for Sia Kangri and Queen Mary for Sia Kangri-II.

Next day Kumar's skis took him to the 20,500 feet high Col where the Camp-III on Sia Kangri had been established. There he caught up with the Advance Party that was preparing an assault on the summit. The sunset there turned out to be the most beautiful he had ever seen. Towering mountains of the western Karakorams stayed silhouetted in the north. Immediately to the west of him rose the Baltoro Kangri or the Golden Throne – and right before them to the north was the famous Conway Col at the head of the Baltoro Glacier. It looked near enough to throw stones at! The only thing that separated him from Conway Col was the head of the Khondus Glacier. There was the point where three major glaciers of the Karakoram met. All of them were touching the slopes of the Sia Kangri massif. The mountains here were so ethereal and the weather so favourably that the scene had a fairy tale touches to it. It was incredible that the same mountains could become so cripplingly dangerous and impassable in foul weather.

But then a storm overtook them and stayed with them up to the 12th of July. Not for a minute was the sky visible. The storm lashed at their tents, obliterated their tracks and pinned them down to their sleeping bags. Visibility in the blizzard was not more than a few yards. White cuts were complete; and worse than the black cuts where at least the eye could get used to the darkness. Whenever the zip of a tent was opened – the snow swept in and swirled the interior, covering everything with ice crystals. At night Bull lay in his tent bearing creaking of tent cords and flapping of nylon flags. Would the nightmare ever end?

By and by their supplies began to dwindle, and finally there came a day when there was not even enough kerosene oil to melt snow and get some drinking water. In desperation they decided to go down to a lower camp and fetch some supplies – especially kerosene oil. People had told him tales of how parties in blizzards had lost their tracks and gone around in circles. But he was not to be deterred as he was clear that it was better to lose the route and die than to die starving in the camp. However, as a precaution he collected all the available ski sticks and poles and also tent poles to be used as route markers on the trip.

While going down they marked the route so carefully and well that even in the foulest of blizzards, they would not have any difficulty in finding their way back. In white outs and in dark jungle one could keep circling around, if the leader's had one foot smaller than the other, as was the case with Bull. They finally made it to the lower camp only to find that they too had no supplies to spare.

They waited for a party from the lower camps to come up, but that party did not show up. They, thus, returned disheartened and disappointed not having found any food to cheer them on their return journey. They, however, collected all the ski sticks and poles they had used as route markers. If the weather had remained bad for another couple of days they would have abandoned their attempt on Sia Kangri entirely. But luckily for them, the weather improved and not only did they received fresh supplies from below; but they were able to push Camp-IV to a height of 22,500 feet on the western slopes of Sia Kangri. The route lay first over an icefall and then over a ledge to the bottom rib that turned out to be the only safe place for their camp.

While the summit party was going up, a moderate sized avalanche came hurtling down. It had broken loose and was heading straight for the eight-member assault and summit party. The few seconds that the avalanche took to roll down the mountain seemed like hours. They waited with abated breath and with their hearts in their mouths in utter despair. From below it looked certain that the party would be completely engulfed. But the avalanche lost its force by the time it reached the party. They only felt the wind blast. The snow debris, about 2 feet thick, stopped a mere couple of feet away from them. As the mountain settled down, they breathed a sigh of relief.

Though Sia Kangri had been climbed once before from the Baltoro side, they were making the first ascent from the Siachen side and therefore had to select their route. The entire mountain face seemed dangerous except for a small rib that took them up to a huge hanging ice wall. They could not fathom as to what lay beyond, for the summit of the mountain was not visible from where they were.

On the fateful day of the 13th of July, the summit party of five members led by Major Chopra – the Deputy Leader of the expedition – left Camp-IV. In the meantime, Bull decided to take the support party up to Camp-IV in order to be as near the summit party as possible in case of an accident. That camp offered a stupendous view of the Baltoro Kangri plateau and its four high points. From there, it seemed that if they were to throw a snow ball northwards, it would breach the Baltoro Glacier. If they were to throw it right in front of them the snow ball would reach the Khondus Glacier and if little to the left it would go into Siachen Glacier. The Karakoram had been called the third pole because of its huge glaciers. If the centre point of that glaciated area had to be selected, that would have to be the summit of the Sia Kangri! Apart from the three glaciers already mentioned, the slopes of Sia Kangri were also connected to the Urdok Glacier of Central Asia.

They looked up the mountain and saw their team make tremendous progress. They had started traversing beneath the ice wall westwards. It was a very steep ice traverse and twice a couple of members slipped and were held by the others. Anybody who fell would have come straight down to the yawning crevasses below. After a time they disappeared behind a crest line.

When the team at the lower camp spotted them later at 4:00 pm, they were certain that they had got to the summit and they made arrangements for their reception. It was only at 6.00 pm that they reached Camp-IV. They were happy to see them all back, safe and sound. After a swift interrogation of their climb, doubts began to arise in Bull's mind if they had climbed the real peak or not. Major Chopra's account further confirmed his doubts. According to their account there was another peak 3 km away from the one they had climbed. That was also six to seven hundred feet higher; but they felt that was one of the Gasherbrum peaks. Bull could not bring himself to agree with them. He decided to keep his reservations and doubts to himself for the night in order to give him some time to think over the course of action to be followed. He, therefore, sent that party down as quickly as possible, as there was not enough place for all of them at Camp. He certainly felt bad pushing them down after the long day, but in the larger interests of the expedition it had to be done.

The whole night he debated with himself whether they should mount another attempt on Sia Kangri or leave the matter where it was. One thing that clearly emerged in his mind was that if the second party had to go up he would certainly have to accompany them because of the doubts that had been created over the real summit of Sia Kangri. He decided to go.

The next morning, though not with much enthusiasm, the climbers laid out a plan of action. They divided themselves into two parties. The first party consisting of Captain Pathania and Havaldar Rana were to leave the camp 45 minutes ahead of the second party. The second party consisted of Subedar Des Raj, Havaldar Vinod Kumar and Bull. There were two reasons behind splitting up of the party into two groups. Firstly, in a small camp it became very difficult to feed five people at the same time in the morning. Secondly, instead of waiting on the slopes for the first two to make the route it was better for the second party to get there fresh and to take over from the first party.

Thanks to the steps already made by the party on the previous day their progress was quicker. What the earlier party had covered in five hours, they were able to do in two hours. In fact Bull's party caught up with the first party just as they reached the wall. There he took a decision to avoid the icy traverse completely and circumvent the wall from the right side. A lot of kicking of steps and crampon climbing had to be done on the last portion. Then they suddenly found themselves on level with the upper portion of the ice fall. There he was surprised to see a huge ice plateau.

On the western end of the plateau was a rounded feature that had looked like a peak from below. On top of that they saw a flag fluttering away. That was the peak climbed by the party of the day before. On the northern end of that plateau was a sharp tooth like peak that to him was undoubtedly Sia Kangri. Towards Northwest they could see the huge towers of the Gasherbrums (I, II and III) – a whale like peak was the highest of them all – Gasherbrum-I. At 26,490 feet, it was only next to K-2 in height in Karakorams.

One of them made a wild guess that it would take them less than an hour to get to the summit from where they stood. Another made a more sensible guess of an hour and a half to the summit. In fact all guesses turned out to be wrong and it took them three hours to get the top of Sia Kangri. To begin with the plateau turned out to be almost three km wide and they were sinking 6"- 8" in the soft snow. When they finally got to the base of the summit, the climbers found it too steep to climb without cutting steps. One of them even slipped 20 metres before he could be stopped. It was already late at 1.00 pm To cut steps for 500-700 feet would have taken too. So one of them took the lead and climbing on points of crampons carried the rope up for 100 feet and anchored it there. Others came up with the help of the rope; that process was repeated six times till they got to the stony portion 30 feet below the summit.

Here they found a flag mast left behind by an earlier expedition, to that they added their ice axe inscribed with the names of Mrs. Indira Gandhi, General KVK Rao and summiteers. As the actual summit was badly corniced, they decided to have their traditional cup of tea and juice on that rocky bit instead of on the summit. After that breather, they all covered the remaining thirty feet to the summit in turns. That was done so that there would always be someone to belay the climbers in case the cornice suddenly gave away. From the summit they could look into Afghanistan, Russia, Chinese Turkistan, Tibet, India and Pakistan. The weather was extremely clear and they were able to take some cine shots from the summit. They also got view of Baltoro Glacier, Khondus Glacier, Urdok Glacier and of course the Siachen Glacier.

The return journey did not turn out to be as pleasant as the journey upwards. They started downwards at 3:30 pm Vinod Kumar was feeling completely dehydrated and exhausted. Once he suddenly slipped for about 20 yards. Seeing that, Bull quickly threw himself on the ice and anchored him with the ice axe. He was thus able to arrest his fall. If he had also gone down with him then it would have been very difficult for the others to save the entire rope. The weather held till they got to the end of the plateau and then a terrible blizzard began to lash at them. Their exhaustion made the slope seem much steeper then than it had appeared during their climb up. So Bull decided to move very slowly belaying at every step. Havaldar Rana and Bull took turns belaying all the way down. Rana would fix an anchor and lower a rope and let all of them go down. Then Bull would fix an anchor while the others went down. They must have repeated that at least 20 times before they reached easier ground. The moment Bull sighted safety all strength ebbed out him. It became sheer agony for him to cover even a few steps at a time. It was at 9:00 pm, engulfed in utter darkness, that they hit Camp-IV.

Major Chopra and the others had watched them from below go up like a bullet but they watched them them coming down at a snail's pace. They were convinced that an accident must have taken place and that they were bringing down a casualty with them.

The most pleasant surprise for them on reaching the camp was the sudden spurt of energy in Havildar Vinod Kumar. Near the summit he had nearly collapsed with sheer exhaustion and dehydration. He had complained of headache and a feeling of nausea. All that must have been only due to the high altitude for he then miraculously recouped. So much so, that Vinod Kumar even made a cup of tea for them before they retired for the night. On the return journey Major Chopra and Bull made a trip to Sia La – 18,500 feet and they had an excellent view of Chogolisa (the Bride Peak) – Baltoro Kangri and Khondus Glacier that turned to the left in a snake like movement.

Before going to Indira Col and Sia Kangri, Bull along with Captain D.K. Duarah made a trip towards the base of Saltoro Kangri on the 23rd of June. Saltoro in Tibetan is "GSL – Gtor –po" the giver of light – perhaps connected with description of the glacier glittering in the sun. That peak was triangulated at the same time as K-2 and was given the number, K-10. Later on, not only was its local name discovered but given to the subsidiary range of the Karakoram that runs southwards from Sia Kangri and divides the Western Karakoram from the Eastern Karakoram. It also gave its name to the pass that was earlier considered to be connected in between Baltistan and Yarkand as the Saltoro Pass (18,200 feet), was also called Bilafond La. After an hour walking up the Siachen Glacier they got the first and the most magnificent view of the peak.

According to Bull, "Mountains like people have their individuality and character, some are always common place no matter how high they tower; but that mountain is so noble in its build – strikingly tall and graceful – and it is so supremely picturesque and beautiful that it is like few commanding personalities, one meets so rarely in life."

From there they turned left and followed the glacier coming down from the twin peaks of Saltoro. The going was excellent and he agreed with an earlier expeditions report that the glacier was the gentlest of all the Siachen's tributaries. After 10 km the glacier ended in an immense flawless, snow expanse encircled by Saltoro, Sherpi Kangri and four more peaks above 22,000 feet. What a feast for climber's eyes! Their eyes were set on the graceful sublime – the set of Saltoro jewels. They traced a line of tentative ascent. The route selected, went south till the base of Saltoro and then up its eastern face. The first obstacle was an ice fall about 2000 feet in height. It appeared dangerous and deadly but then there was no other alternative. That route met the south ridge at about 23,500 feet and then followed the ridge to the summit. Having satisfied themselves that the route could 'Go' they started their return journey.

Considering their rapid progress on the way up, they thought that the descent would not take them more than couple of hours; but that so called gentle and placid glacier, having warmed itself up, was ready to show different colours. While going up the entire surface of the glacier was frozen solid and they did not even imagine what lay under that harmless looking snow cover, till Captain DK Duarah walking right in front of Bull disappeared in a snow hole. His hands and ski sticks were the only parts of his body that were visible. One could well imagine Bull's consternation. He found it hard to believe his eyes. Mehrwan went to extricate Captain DK Duarah and got trapped himself there. There was danger ahead and they all fell on their fours. Luckily the crevasse was not deep. Kalam Singh crawled up and helped Captain Duarah and Mehrwan out while Bull held his leg.

It was an exhausting job as wherever they put their feet, they sank thigh deep. The top layer of snow having become soft with heat of the sun gave no protection against the yawning crevasses, as it did in the early hours of the morning. That was only the beginning. Then onwards, every ten steps or so they fell into the crevasses that luckily were not deep enough to swallow a man, but could break their legs at any time. At no time could Bull see more than two people out of snow. In a way it was not climbing but wrestling with snow. After 6 hours of gruelling effort they returned to Advance Base Camp totally exhausted; but they were lucky to be spared any casualty.

As the weeks passed these crevasses opened out; and at a later stage one of the members Mangla Rai fell into a 30 feet deep crevasse and was precariously balancing on a snow 'chalk stone' while the gushing cold water stream flowed only two feet below him. On another occasion, Ved Prakash and Rajinder Singh were carrying loads from ABC to Camp-I. They had gone for 2 km when Rajinder Singh looked back and found his companion had suddenly disappeared. He stood paralyzed for few seconds and then returned to see the gaping hole that had engulfed his partner. He kept yelling into the crevasse, "Are you there, are you there?"

After some time a feeble voice was heard from great depths of the crevasse. He was relieved to get a response and rushed back to ABC for help. Ved Prakash was inside deep freeze 75 feet below for two hours before he was evacuated, with only a broken ankle. That was not a gentle glacier but the very devil's glacier full of death traps to entice unwary men into its pitiless jaws. Surprisingly, the longest western affluent of Siachen Glacier (10 miles long) had no name. So we decided to name it Saltoro Glacier after the peaks at the head of that river of ice.

On the 30th of June, Camp-I was established at 17,600 feet at a place little short of the point reached by them earlier. It was called Suraj. On the next day Camp-II was established at 19,000 feet at the foot of the East face, it was called Hema. That camp gave them the inside views of the Saltoro amphitheater. Towards the west was the towering monarch of the Saltoro range, they had to almost sprain their necks to look at the Saltoro Summit. In the north, two 24,000 feet high peaks seemed to challenge them and towards the south was the gap between Likah Glacier -- South of Gong Gong Glacier -- and Saltoro Glacier. Lord Hunt had used that gap to come from Khapalu in 1935. For them that gap was the bringer of bad clouds. Lord Hunt's expedition had to turn back when they were only 800 meters from the summit due to highly dangerous avalanches. The eastern view was abstracted by low lying mountains very close to the camp.

An overhanging ice wall that kept on breaking repeatedly obstructed the route up the eastern face of the mountain. But there was a 15 meter wide gap – a weakness in the mountain from cornices at the end of a steep slope! That took one to the higher plateaus of ice that was about 300 meter high. On that ice field they put their Camp-III at the height of 21,000 feet on 8th July. Rope was fixed on the cornice to help the people carrying heavy loads.

There onwards they had to really battle their way up the mountain. The first party fixed six 200 feet long ropes to overcome the lower portion of the ice fall and when they had got up a ledge half way up, a huge ice avalanche broke just south of them and headed for Camp-III. The entire slope was covered with avalanche cloud and they thought Camp-III was done for. When the cloud lifted, they saw Premjit Lal, Kalam Singh and Thandup running for their lives not wasting their time even to look back even after the avalanche had halted. It was touch and go, as huge ice boulders stopped only 100 meters short of the camp.

The remainder portion of the upper ice fall like the lower one had only one opening and that was through a steep snow gully that seemed extremely risky. The party consisting Captain DK Duarah, Havaldar Kanshi Ram, and Sharma pushed the route through that risky gully and put Camp-IV at the height of 22,500 feet. On the 13th of July, 10 members occupied Camp-IV above the gully and thought of making the summit attempt from that camp itself. However, at night the altitude struck and two of the toughest fell sick and vomited. Negi vomited some blood. The sick were sent down and the party decided to respect the altitude and they put another camp just below the ridge at the height of approx 23,400 feet. There were no technical difficulties from

Camp-IV to Camp-V, however dangers of avalanches could not be ruled out. That Camp was occupied by Captain D.K. Duarah, Havaldar Kalam Singh, Captain Premjit Lal and Havaldar Kanshi Ram.

The next day Havaldar Kanshi Ram got severe chest pain and came down. The other three attempted the summit. After a while Premjit got sick and returned. In the mean time Rattan, Sonam and Subhash moved to Camp-V encouraged by Captain Paramjit Lal who was coming down. They decided to make use of the beaten track and go for the summit. There were no technical difficulties and they needed just ploughing through the soft snow. Kalam Singh broke the trail most of the time. Enroute they came across a huge rock tower that they kept to their left. They kept on going till 3 pm but summit was not in sight. The lateness of the hour and sheer exhaustion compelled them to return when they were approx 300 feet below the summit. Captain DK Duarah who had carried his skis all the way up skied down to Camp-V. It had been a tremendous achievement.

Bull joined that party at Camp-II on the 17th of July. He found the team in bad shape. They had got a beating from the mountain. Most of them complained of exhaustion, severe chest pains and some had vomited their guts out; but their morale was high and Bull knew with a little rest, they would be ready again. The 18th of July manifested bad weather and they discussed the climb that had been abortive and made plans for a new attempt. On the 19th of July a party occupied Camp-III. That night it snowed heavily.

On the 20th morning when the summit party tuned in their wireless set there was a commotion in the mess tent. People from higher up seemed shaken and were yelling, "A huge ice avalanche fell last night and some tents were uprooted with the blast!" There was no doubt that the blast had been a terrible one, as thousands of tons of ice had fallen for couple of thousand feet. The uprooting of tents was a slight exaggeration. Thank God, the camp was safe. They shifted it a little further away from the tumbling ice fall. Camp-IV and Camp-V were also safe but their route high up had been completely obliterated and ropes gone with the wind. The 'Risky Gully' had badly avalanched and became "A suicide Gully". But the Army humour typically displayed optimism. Thandup said "Good now the crevasses are covered with the avalanche debris, so they don't have to bridge them."

"And look at the sheer ice face, no more bogging in soft snow, clear cut climbing," said another climber.

In his twenty years of climbing, Bull had never seen snow volcanoes. He saw two of those then. Those were huge holes in the ice wall through which every now and then snow spurted out in the form of frozen lava. And one of these was very close to their fixed rope route. That was one ice wall he prayed should come down quickly. If that broke off, the route would be safe. One evening they heard deafening thunder. He peeped out of his tent and saw that the entire ice wall seemed to be peeling off the mountain. It was in a slow motion, it looked supernatural and awe inspiring. The huge overhang like cornice curled over. Vast masses of ice toppled over, leaping in thin air rolling over each other, smashing themselves against the rock with hissing, growling and crashing sounds, as if the mountain was venting its anger; and then the ice turned into a cloud, if by magic. The cloud like an atomic blast shot across the slope, regardless of undulations and gradient, making a noise that might be produced by many trains running parallel to each other. The cloud – snow dust- travelled for another 2 km before it halted. Bull was happy as that was the avalanche they wanted and needed so badly. However, when he looked back on the mountain he was horror struck – only half the wall had come off, the other half was even more precariously balanced then, than before! Now they had to wait till the remainder of the wall broke off!

The mountain was playing with them, just as the cat played with mice - testing their wits, patience and perseverance. It was like playing a game of chess with the mountain, in that the mortal pawns were 'Us' and the 'King' that could never be killed - the mountain. He knew then they were beaten. He wished he had mortars to bring down the rest of the ice wall.

That night he slept early and well. It was the sleep of a beaten man reconciled to his fate. From that sound sleep, he was woken up by a thunderous sound-his tent was flapping vigorously. His first reaction was of feeling nonplussed and then he volunteered a smile. It was pitch dark -- but he knew the second half of the ice wall had come off. He was absolutely certain! At least the mountain had fully spent its first line of defence. That was one avalanche they all celebrated.

Then they remade the route to Camp-IV and it was occupied by three members. Another four were to join them on the 23rd of July. At about 10:30 am when the second party was half way up, he picked up some snow and found it wet. The night's snow fall had been powder dry and would that stand the weight of new wet heavy snow? He had his doubts. He tuned in his wireless at the next 'call' and called everybody off the mountain. They tried to protest but he was firm – very firm. On return they cursed like sailors! It was his intuition or sixth sense or whatever one wished to call it. That night the entire mountain peeled off. Camp-IV where 7 people would have slept that night was under 200 feet of snow and absolutely nothing could be recovered. What was worse - the second party had left all the cooking gas and that was also lost. He knew, however, if there was no more heavy snow fall the 'Mountain' had spent its second round of defence.

On the 31st of July Camp-III was reoccupied and on the 1st of August 1981 the summit and support team once again passed through the ice fall and suicide gully and went straight to Camp-V skipping Camp-IV site. That was indeed a very long day but also safer. Only once the support party returned after leaving the four of them at Camp-V, did Kumar feel relieved. There onwards there was very little danger due to objective hazards – but if someone fell ill, it was a different matter.

On the 2nd of August the weather dawned clear but as usual the cloud hung over the top portion of Saltoro peaks. They saw the summit party consisting of Kalam Singh, Tendup, Gaj Bahadur and Swarn go up. Kalam Singh was climbing the same peak, for the second time. Two weeks earlier he had missed the summit by few hundred feet. They headed north, traversing the main slope, and then they were out of site behind the snow knoll. Bull took an air mattress, wireless set and some tea and took position on a snowfield opposite Saltoro. It was very hot and unbearable. So after two hours, he returned to the shelter of the mess tent. They could do nothing except wait, and that was a long painful wait indeed!

Only at 6:00 pm was the party seen again returning to Camp-V and at 7:00 pm all were in, safe and sound. Kalam's voice was jubilant, they had done it. They reached the summit at 2:45 pm and hoisted two flags on either side of the summit. The eastern side was a little clear and they took pictures of Sherpigang Glacier. They could see some villages in Khondus valley. Next morning when the mist cleared one of the Jawans fixed spot scope. Bull thought he was focusing on the camp till he shouted "the flag – the flag". Bull looked through that and could clearly see the "flag" fluttering away in the wind on the highest point in the eastern Karakorams. Before leaving the mountain, they skied up to the pass south of Saltoro that lead to Sherpigang Glacier, and western Karakorams.

It is said, "Mountain is like a sea, an exacting mistress, but ever ready to give her favours to those who woo her with great devotion." At last the Bull and the Bull's team was awarded its dues after Herculean climbing against heavy odds.

KAMET - 1983

In 1983 the time came for Col Bull Kumar Bull to say farewell to the Army. He had been in uniform for almost 34 years. While Bull was preparing his retirement plans that needed all his attention as one retired at comparatively young age of 50 in those days. He got a message from the then Colonel of the Kumaon Regiment Lt Gen PN Kathpalia, VSM, AVSM, PVSM expressing hope that before he laid down the arms, "He should lead the Kumaon Regiment Mountaineering Expedition so that the Regiment had some experienced climbers to pursue adventure activity bringing to the fore qualities of the leadership and courage to face severe adversity of the unforgiving mountain climbs".

Bull being a die hard Kumaoni indeed ~~is~~ was very proud to be one of the clan. Ask a soldier, almost any soldier, and you will find him utterly, even fiercely and aggressively loyal to his regiment. To a soldier, his regiment is his second home. It is what nurtures and shapes him, gives him a sense of identity with a group of fellow-soldiers, gives him proud traditions to carry on, and a heroic and glorious past as legacy. So Bull just could not say no to the desire of the Colonel of the Regiment but the question was the selection of the mountain to climb. It had to be high enough to get recognition, not too difficult to climb as most of the Kumaonis would be the first timers He, therefore, selected Kamet, the third highest mountain in the India.

He had already successfully led expeditions to the highest two; the Kanchenjunga and the Nanda Devi. So that could be a satisfying hat-trick. With a little more enthusiasm, he then turned to the earlier accounts of Kamet to crystalize his Regiment's aspirations.

It was first climbed by Frank Smythe's team in 1931, the highest mountain ever climbed till then. A continuous series of expeditions thereafter set out. With India's independence, Indian climbers too found in Kamet a friendly and attractive mountain and were successful in climbing it in 1955, 1970, 1980 and 1982. All these climbs were made from the classic route- the Eastern Kamet Glacier followed by Smythe.

Surely there were other aspects to a mountain as high as Kamet. He looked for photographs, studied them, and gradually a vision began to form in his mind. He discovered that no one had attempted to climb the mountain from its western flank since the intrepid Captain Slingsby's unsuccessful efforts of 1913. He just could not believe that. Here lay the highest unclimbed route on any Indian mountain, and none had thought of it for over seventy years. He decided to go up and take a look at that baffling mystery.

The primary goal was, of course, to initiate the Regiment into serious mountaineering. Kamet at about 25,447 feet (7756 m) would certainly be a good training. A successful ascent would generate confidence and get Kumaoni climbing activities off to a good start. Then, considering that substantial commitment of effort and money would be involved in the venture, they decided to make the most of it. The regimental mountaineers, thus determined to make a full-scale reconnaissance of the Western flank of Kamet, an approach never before seriously tackled. He had no problems in organising the expedition as Brigadier Satish Issar the Commandant of the Kumaon Regimental Centre, took upon himself to look after the logistics; the most complex task requiring minutely detailed preparation for any Himalayan expedition!

On the 30th of May, he drove up from Joshimath to Mana Camp at the road head, a little way above Badrinath. Even before Badrinath, his eyes were inevitably drawn to the majestic Nilkantha, the mountain of such tragedy and triumph for his small team in 1961. From Mana Camp to Ghastoli was approximately 16 km that they covered the next day. The route lay over the frozen Saraswati River. The 3,300 ft climb took them 7 hours, a nice testing walk for starters. From the Base Camp at Ghastoli the boulder strewn terminal moraine of the Western

Kamet Glacier was about a km's walk along the left bank of Saraswati. The glacier itself was almost exactly at right angles to the Saraswati, a ninety degree turn to the right as one walked up from Ghastoli.

Out of their 50 porters, 30 of them revolted after going on the snow covered route. Dhotials from Nepal were capable of carrying almost 60 kgs on the back over steep climbs but the first encounter with snow and they were gone. The path climbed in a steep zig zag and spread around a high shoulder. After a km's climb the gorge opened up and the route then went over the grassy and pleasant hill side, after 3 hrs' walk they reached the Base Camp situated under the rock cliff that was a tiny green patch at the height of 13,700 ft. Next to that they found a fresh water spring. The route from Base Camp to Camp-I (16,500 ft) was on lateral moraine and tough for green horns and veterans like Bull.

On the 4th of June, he walked over the snow track more easily than the day before. The early start certainly made things easier along with his growing fitness. Mana peak was visible but they had to wait for the grand view of Kamet. After an hour's walk they saw the formidable southwest face of the Kamet and the west ridge. They could not yet see the complete ridge line leading to the summit. But they could see enough to realize that the going was to be very tough. Some routes available were threatened with avalanches, some routes did not have enough camping places and some were sheer.

At Camp-II they met Captain Suraj Bhan Dalal, 2 Kumaon (Berar), Deputy Leader in charge of the western route. He had served as the instructor at the High Altitude Warfare School and was member of the Sickle Moon Expedition-the highest mountain in the Kistwar Himalayas. He had spent the last three days considering the various possible climbing routes, gauging the risks and dangers, searching for possible camp sites, calculating the logistical build-up needed to propel the team on to the proud high point of Kamet that rose 7,000 ft above Camp-II. For detailed reconnaissance they walked to the rock face, the first obstacle before gaining the crest of west ridge from southwest side. The rock step wore sheen of verglas with very uncertain holds. Worse the gully above that they appreciated to offer a line of crest that was much steeper and was a natural funnel for the falling stones. There was no camping place or even a resting place. And three thousand feet of climb in one stretch had to be managed; a feat that was possible but very difficult and a dangerous proposition!

They then turned their attention to the southeast ridge. To reach the high Col, they first had to pick a their way gingerly through a highly crevassed glacier -- a long detour traversing the lower slopes of Mana peak well beneath higher reaches, and then a final steep ice wall to the Col. That latter section would again be exposed. From the Col the southeast ridge rose steeply at an angle of over 45 degrees -- a sharp challenge at that altitude. About a thousand feet below the summit that turned into an overhang of about 100 feet which they over came by fixing the rope but at 24,000 ft it would be a difficult to execute. Having observed all these obstacles they returned to Camp-II.

The next morning they decided to examine the approach from the northwest face. After a 2 km walk, they turned right and entered a small closed valley about 2 km long that ended in a high ridge that seemed to link Kamet on the right and Mukut on the left. Going into the closed valley, Kamet lay to their right, its northwest face rearing upwards for over 6000 ft. On the left they could see bulk of Mukut Parbat with a network of broken ribs in its upper bastions. The valley stretched as an almost flat snowfield till that began bumping into huge ice blocks and avalanche debris that extended for about 200 yards from the base of Kamet. Then they looked up, their necks craned as far as possible, their eyes focused on to the greatly fore-shortened view of Kamet's northwest face looking for a gully or a rib that might offer a lead to its lower slopes. There were no such obvious climbing routes. The lower half of the face first rose almost vertically upward in avalanche ice cones and then turned into a dark rock wall around relieved by the whiteness that on closer observation was found to be a flake of thin ice.

At about the height of 23,000 ft, the entire face was girdled with flaring hanging glaciers. Three of them stretched right across the huge expanse and it did not require any imagination to conclude that their blue undersides - seemingly unsupported - would intermittently sag and break off. There were three of those, each looking nastier than the others. Above those hanging glaciers the going seemed easier.

Thus, one could reach the Col on the west ridge about 1,500 ft below the summit from where they did not see any major obstacle after the glaciers.

During his earlier reconnaissance Suraj had zeroed in on one of the possible routes between the last two glaciers that seemed safe unless an abnormally large portion broke off from the left hand side of the central glacier. It's over flowing debris could threaten about 500 ft of climbers' route. They asked themselves -- work early at dawn or climb at night with head lamps? They mulled over the options as they continued to scan the line up the face. After protracted discussion amongst Suraj, Bull and Naib Subedar Kura Ram, who had been a member of Bull's Kanchenjunga Expedition, they decided in favour of the northwest face. Actually it was decision between the devil and the deep sea. Suraj seemed very happy as he thought that the only feasible route. The reconnaissance party was also carrying some loads for Camp-II that they dumped at 19,200 ft that included an orange Mead tent named after the same person that carried the name of Meads Col- the gateway to the summit of Kamet.

The first route opening party on the northwest side consisted of Mao, Rajendra and Balwan. It took them about 20 min to get to the base of the avalanche cone from Camp-III. After cutting steps for 700 ft, they returned to the camp at 3:30 pm. Mao's arms and shoulders ached due to the endless swinging of the ice axe - he had done most of the step cuttings. After a day's rest they went up again. Balwan had to go down as he was not feeling too well and his place was taken over by Lakhpa, a Sherpa from Darjeeling. They followed the earlier footsteps and fixed some more rope ahead of it before returning to Camp-III.

On the 9th of June a rope consisting of Kura Ram, Dan Singh and Abhay decided to take on the northwestern face. An hour and a half later they crossed the 20,000 ft mark and hit the high point reached by the earlier rope. Immediately, their progress slowed down as they were moving one at a time alternately over rock and ice. The face offered very few secured holes and artificial ones had to be created. That day Kura Rams team fixed another 600 ft of rope gaining additional 300 ft on the mountain.

As the main objective of their expedition was to train Kumaonis in high altitude climbing and a majority of them were on the eastern axis that was an easier and better route, Bull decided to go on to the eastern side with them. On the eastern side their Deputy Leader Captain DB Thapa was in charge. He had already climbed Kamet with the IMA expedition in 1982, and was very competent and remarkably mature. He followed the same route traversed by Smythe -- the western glacier through a rock gully to Camp-III below the Mead's Col.

The eastern party got involved in a serious evacuation of Lance Naik Bharat Singh, who suffered from pulmonary oedema at Camp-II and was carried down to Base Camp. A strong mountaineer could climb the summit of Kamet from the Meads Col camp at the height of 23,500 ft. Captain Thapa suggested that another camp be put in between Meads Col and the summit. Bull agreed hoping that it would be also act as a security camp in case they crossed over from the western side, a wishful thinking. A Summit party of Captain Thapa, R.K. and Balwan made an attempt on the 24th of June, but they fell short of summit by 500 ft.

On the western side Captain Sambial's party going from Camp-II to Camp-III had a harrowing experience when Pratap fell down into a crevasse. Looking downwards he saw 100 ft walls of blue ice, and he himself was 10 ft below the lip of the crevasse. However, the rope helped him come out of that frozen hell.

On the 10th of June, Kura Ram's team set out early at 10:00 am and they reached the previous high point in two hours' time. They were stuck at a chimney that did not look very promising. Dan Singh who was watching Kura Ram's anxious face decided to tackle the left edge of the rock wall that was icy. The ice was too thin to put on pitons had no protuberances, no hold at all, in fact nothing! At one place Dhan Singh climbed on Kura Ram's shoulders to hammer in the pitons. Having got over that holdless pitch, there was no stopping him. He climbed for another one hour fixing more pitons.

Down below at Camp-II they watched these rock manoeuvres of Dan Singh with fingers crossed. Later Major Ravinder Nath said, "He went up as if the summit was insight, swift, surefooted and smooth;

he was a man inspired." He wanted to do a little more that day but his senior colleague Kura Ram thought that they have had enough for the day and came down. Suraj Dalal reckoned that Dan Singh had definitely climbed over 20,000 ft. From the 11th of June a spell of bad weather set in that carried on for a week. Captain Sambial and Captain Aggarwal attempted the route but could not get to the point reached by Dan Singh.

On the 19th of June Suraj sent up Dan Singh and Lakhpa. It took them five hours to Jumar up the avalanche cone, rock wall-one, than the 500 ft of ice slopes and the large boulder at the foot of rock wall-II, the previous high point of the team. Rock wall-II grew over 150 ft in verticals and overhang and nowhere did it soften to an angle less than 80 degrees. It was a heart breaking obstacle to encounter after a strenuous 5 hours' ascent; but the pair had made an early start and weather held and they were able to fix the rope on rock wall-2 in about 3 hrs of exceptional climbing. They arrived back at Camp-III at 5.00 pm in high spirits but unfortunately they could not find the place for Camp-IV.

On the left of the rock wall-2, and separated by 40 ft of ice slope, was another rock wall of the same size and steepness as the former. The top of that rock wall should be approximately 22,000 ft and above that was a steep slope of approximately 300 ft that would end at a point above the hanging glacier, through a tongue of hard snow. From the ice slope to the glacier would be demanding 500 yards traverse but that was safe from avalanches and once they got to the glacier there was a satisfactory camping place. Whether that camp could be stocked with supplies; and the summit could be climbed after putting another camp at the west Col was a formidable question.

On the 21st of June Dan Singh and Lakhpa got over the surmounted and fixed rope on the half of rock wall-3. However, until then no camp space was available where one could bivouac in a hammock; but those Alpine style tricks did not work in the Himalayas.

On the 26th of June a party of 5 - Captain Dalal, Dan Singh, Lakhpa, Rajinder and Abhay Singh started for the face. The bold plan was to reach the Camp-IV site, spend a night there and then go to the summit. A support party of two - Sher Singh and Bharat Singh would bring up the tent and other supplies. When they were at a ledge just below the rock-3 it started snowing. As the tent had not fetched up they decided to bivouac at the base of the rock, belaying by ropes and pitons. There were only three sleeping bags between five of them and they had to share. The night was arduous. The windswept snow was playing havoc with the climbers. Two complained of numb toes and the thought of frost bite was alarming!

The morning dawned bleak, weather remained hostile and there was no hope. Staying there was useless. Dalal was worried about Rajinder Singh and Abhay's numb feet and took a very apt decision to return. His own right hand became numb and he had to do a lot of massaging to bring that back to life. Between rock wall-3 and 2 a small powder avalanche buried Rajinder with the snow. Lakhpa and Dalal also came down very fast. The ropes could have broken, the pitons could have come out and the sling could have deceived them. Many fatal mishaps could have chanced to occur but mercifully they were spared. Again at rock wall-1, they were subjected to a powder snow avalanche. Luckily Rajinder and Dalal were at the bottom of the wall when the avalanche hit. Dan Singh was thrown from the top to bottom of the rock wall-1. It had taken him hours to negotiate that but he came down in a few seconds.

The avalanche passed, visibility improved and one by one they emerged from the snow. Luckily nobody was hurt. At 7:30 pm in the evening they reached Camp-III. The occupants of Camp-III were so worried for the climbers who had spent a night out at an immense altitude in vicious weather, that they could not control tears of joy. Captain Dalal and his men had done an excellent job; they had started a new trend for Indian climbing. Few Indian climbers or expeditions had attempted big mountain faces. Following their southwest route that they had rejected, an Indo-French team got to the summit from the west ridge.

On the 30th of June, Captain Thapa with a seven member team including Lieutenant Bakshi and Lieutenant RK Singh left Camp-II and reached Camp-IV. But they were shocked that all the gear including the tents was buried under snow and what was more the gas cartridge had leaked. Their stove was not to be found either. They had a tough time beating the trek once again after the snowfall. They were very tired and went straight to sleep. In the morning they could not see their tent at Camp-V and had no option but to start for the summit from Camp-IV. They left at midnight taking turns and beating the trek that was almost knee deep in snow. They fixed rope of about 200 ft on a vertical slope full of ice. They reached the summit at 4.00 pm and returned to Camp-V after 15 hours of gruelling effort.

They had been lucky on the eastern side as they just made the summit before monsoons started. The task was accomplished and the expedition had produced some seasoned climbers. Unfortunately one of them, Lieutenant Ramnik Singh Bakshi who had climbed both Kamet and Abi Gamin with their expedition went with an ill fated Indian Army expedition in 1985 and never returned.

In an article 'Peak of Endurance' Suman Dubey has stated that every member wanted to go up but there comes a time when one decides that one can not do it due to high winds, storms and blizzards, soft snow and biting cold temperatures that proved more than a match for the Indian Army Expedition which called off its attempt on Mt Everest on November 5 after it became clear that time, energy and luck had finally run out on it.

The expedition was led by Brig Jagjit Singh who was inexperienced to lead Everest expedition and tried to run it more like a military exercise ignoring safety of team members. On October 7, Major Kiran Inder Kumar (younger brother of Bull) fell to his death from above the South Col (7.986 meters). From the account of those who were up on the South Col with them, it is clear that of the four others who died, Major Jai Bahuguna and Lieutenant MUB Rao were equally fit like Lieutenant RS Bakshi and Captain VS Negi. They could have made it down on their own but chose not to abandon their companions in distress. Staying with Bakshi, said the then Director- General of Military Training, Lieutenant-General SR Rodrigues, "was in the best tradition of the army." Having lost his brother Harish, who was abandoned on the West ridge on the 1971 International Everest Expedition, Bahuguna was unlikely to leave a friend to a similar fate. The Army and the Kumaon Regiment lost some excellent young climbers in this unfortunate mishap in an ill led expedition.

SIACHEN DISPUTE: INDIA AND PAKISTAN'S GLACIAL FIGHT

On 13 April 1984, Indian troops snatched control of the Siachen glacier in northern Kashmir, narrowly beating Pakistan.

Thirty years later, the two sides remain locked in a standoff, but the Indian Army mountaineer who inspired the operation says his country must hang on whatever the cost.

Virtually hidden from public view, the world's highest conflict is moving into its fourth decade.

The struggle between India and Pakistan over the Siachen Glacier has even spawned a new term: "oropolitics", or mountaineering with a political goal.

High-altitude war

Derived from the Greek for mountain, Indian Army Colonel Narinder Kumar can justly claim to be its modern father, because his pioneering explorations paved the way for India to take the glacier in early 1984.

But what started as a battle with crampons and climbing rope has turned into high-altitude trench warfare, with the two rival armies frozen - often literally - in pretty much the same positions as 30 years ago.

The vast majority of the estimated 2,700 Indian and Pakistani troop deaths have not been due to combat but avalanches, exposure and altitude sickness caused by the thin, oxygen-depleted air.

"It's been a shocking waste of men and money", says a former senior Indian Army Officer and Siachen veteran.

"A struggle of two bald men over a comb" is the verdict of Stephen Cohen, a US specialist on South Asia, dismissing the Siachen as "not militarily important".

That would perhaps be comforting if the two combatants did not both have nuclear weapons.

Surrounded by photographs and memorabilia of his climbing exploits, Colonel Bull Kumar, now in his 80s, says the struggle was critical to preventing Pakistani encroachment into northern Kashmir.

As with so many long-running conflicts, it began with an undefined border.

THE MERCURY HIMALAYAN EXPLORATIONS

After his retirement Narinder started an adventure travel company called Kashmir Mountain Travel. They started handling tours in Kashmir for Mountain Travel USA. The Mountain Travel USA was having problems handling arrangements in Delhi for the groups and also in the handling of extension tours of clients and suggested that Narinder should work with Mercury Travels. Mercury Travels is a one-stop company of the Oberoi Groups of Hotels that offered the complete range of travel related services and desired to start the adventure travel module increasingly becoming popular.

After the company was formed he was asked to fix his remuneration. He decided not to draw any for the first 6 months as he was getting his leave pending retirement pay from the government. After 6 months he started paying himself just Rs. 3,000 per month, the same amount as the driver of Mercury Travels. His aim was to build up the company for the long run and he had to sacrifice the short term gains.

During the early days, his job was a cushy one. Those days all the international airlines would give travel agency executives tickets at just 10% of the total fare. Mercury Travels had its offices in London, Frankfurt, USA, and Spain and had global contacts. He often used to make trips to various countries on business.

They also used to attend International Tourism conventions like ITB in Berlin and WTM in London. In those fairs, India would hire a huge space and every travel agent had a small office. The concerned India Tourism Office in the city would have a small restaurant from where one could order tea, coffee, beer, wine, etc. They would meet about 30 travel agents from various countries abroad and try to sell their products. Occasionally, every country used to have happy hours in the evening where the drinking started at 5.00 pm in the evening, after that there used to be cocktails at some country's pavilion followed by dinner by the tourist departments of the various countries. The hazard of that life was that they had to really guard against one's drinking as one had to wake up early in the morning to go to the fair and open a stall and start working again.

The programs were very hectic and they had no time to do any sightseeing of the country they visited. It was very difficult to break into an adventure travel market. But, their success story was that they offered treks that other adventure travels did not like taking the American groups to Indira Col – the source of Siachen Glacier. They were first to start a jeep safari from Manali to Leh.

When their principal abroad published a new itinerary, it was a hit. The jeep safari trips were oversubscribed and they decided to start one trip from Manali and another from Leh. They met halfway, changed the vehicle and completed the trip. They were also the first one to go down the Ganga River from source to sea. Though it was a 45 days trip, it was also oversubscribed by a few persons. They had to refuse some of them as they had to build a new boat to take the group down from Haridwar to Calcutta and Sundarbans. It would have been expensive to bring back the boat, so Bull Kumar decided to give it to the Eastern Army Commander Lieutenant General KS Brar, MVC of the Blue Star fame. The funniest part was that, their small boat carrying 16 people was escorted by a big ship in the Hoogli. The Americans were very happy to get such warm reception. The trip was repeated again for the National Cadet Corps cadets, but except for the crew, the passengers kept changing. As the trekking trip around Annapurna was even more popular than the trip to the Everest Base Camp. Bull Kumar decided to start a trip around Nanda Devi too. His first trip was very successful.

However, for security reasons, it was discontinued. The situation was a very awkward as the bureaucrats sitting in the air-conditioned offices did not understand the ground realities. One could go up to Ulta Dhura from the Munshiari side and Kingru Dhura from the Joshimath side but that small gap of not more than half a km was the so called 'restricted area' and foreigners were not allowed to go there. There were tremendous opportunities for adventure travel in India, but due to restricted area conditions, the adventure tour operators could not market that.

Another example that came to Narinder's mind was rafting down the Brahmaputra from Tuting – the place where Brahmaputra entered India to Pashighat -- the restricted area. Narinder had organized the first Indo-Japanese Expedition down the river with special permission. Narinder felt that if India kept having restrictions and did not allow foreigners till Arunachal up to the Tibet border, China would fortify its claim over that territory as the so called Southern Tibet. A long time ago Chow-en-lai, the Chinese premier had made an offer to Prime Minister Pandit Jawaharlal Nehru that China was prepared to give up its claim on Arunachal Pradesh if India gave up its claim on the Aksai Chin area that they had occupied and through which they had made a highway in 1957. In that connection, Narinder quoted an article "India and China need to make a fresh beginning in bilateral relations" that was published in the Times of India 0n the 24th of March 2014.

Earlier Penang Tso Lake was restricted to foreign tourists. With great difficulty that was got open. There were many other areas that could be usefully utilized for adventure tourism purposes. Bureaucrats of three ministries – Tourism, Home and Defence have to sit together to sort out nagging problems. Adventure Tour Operators Association (ATOA) should take up that case strongly with the Ministry of Tourism.

Bull Kumar remembers he had arranged an Indo-Dutch expedition to Base Camp of Gorichan peak. The Indian Army Officer was the leader of the expedition and they also had a liaison officer from the Indian Army in the area. The group seemed to have crossed one day before the arrival of the sanction from the Intelligence Bureau (IB) in that area and Narinder's company was blacklisted. Thanks to Lieutenant General Lakhera, who was then the Adjutant General at

Army Headquarters and Colonel of the Kumaon Regiment and Bull Kumar asked him to help Indian Army in starting Adventure Travel Company for the Army Wives Welfare Association (AWWA). It was an Army expedition and he had to pay a heavy price for the same till General Lakhera personally spoke to Director General Tourism and the case was finally closed.

Since the inception of Narinder's company, it has not made any loss till date. It also owned a small hotel and a beach on both sides of the Ganga River. The company has won the government's **'Best Adventure Tour Operator Award'** twice in a row, for promoting adventure sport, an unprecedented feat. At the time of writing this biography, Narinder's son has taken over the company, but Narinder has still retained the authority to sign the cheques with him.

THE INDIAN OLYMPIC ASSOCIATION

Narinder had joined the Olympic movement in 1970 when he was the Principal of National Ski School. One of his students Shri RK Gupta had started the Delhi Ski Association and asked him to become President. Narinder readily agreed as his aim was to popularize skiing in India. The Winter Games Olympic movement was started by Raja Balinder Singh of Patiala whose son Randhir Singh had been the Secretary General for ages. Shri KP Singh Deo, who was the Minister of State (Defence), Government of India, was made the first President of Winter Games Federation of India (WGFI).

However, after a few months, he decided to give up WGFI as he was also the President of the Rowing Federation of India. He transferred his responsibilities to Vice Admiral Jain, who was head of the Services Sport Control Board and posted in Bombay as the Head of the Western Naval Command. During his command, the head of the Ice Hockey Federation visited India, and Admiral Jain treated him aboard a ship. He was very happy and gave WGFI an affiliation to World Ice Hockey Federation.

The first National Championship of Ice Hockey was held in Leh, Ladakh in February 1990. There was a pond there that freezes during winter and people play ice hockey during that time. Ice Hockey could be called the regional game of the Ladakhis. During winters all the passes to Ladakh are closed, and wherever water surfaces freeze Ice Hockey is played.

Unfortunately, the infrastructure for Ice Hockey had not been developed at all. The SAF hold a championship every year in different countries like Pakistan, Sri Lanka, Bangladesh, Nepal, etc. Once while Narinder was the President of WGFI, he flew to Colombo and with the help of Raja Randhir Singh, the then Secretary General decided to hold the SAF games in Winter and Narinder bet India should be the first country to hold those games. India was not prepared with the infrastructure and the games took a long time to materialize; but whenever such games were held in any country, the government was duty bound to provide infrastructure. For example, during the Commonwealth Games, the infrastructure of Delhi improved tremendously. Though there were some charges of corruption that gave a bad name to the Commonwealth Games, those were the best organized Commonwealth Games ever held in India.

The SAF games were assigned to Uttara Khand and the State government and sports ministry constantly pressurized the State Government and Sport Ministry in providing the required infrastructure. There was no doubt that the ideal place to hold any Winter Games was in Gulmarg, Kashmir. Snow slopes in Gulmarg are as good as the snow slopes in Switzerland or any European Alpine country. But unfortunately, due to the political tension between India and Pakistan, Pakistan would never allow any SAF games to be held there.

When Narinder started the National Ski School in 1970, there was only one small T-bar lift. But now they have a Chair lift and Gondola that goes up to 9,000 to 14,000 feet. They also have snow beaters to pack the slopes and the snow ploughs to open the snow covered road from Tangmarg to Gulmarg. In those days all small hotels used to close down during the winter. He remembers giving subsidy of Rs 50,000 to Highland Park Hotel to keep that open during the winter. After some time the hotel decided to do away with the subsidy and charged its own rates. At present, there is the Khyber Hotel – a 5 star property that has good occupancy during the summers as well as the winters.

To provide infrastructure the Central Government gave a considerable amount of money to the State Government with the result that there was a first class ice skating/hockey stadium at Dehradun where Ice Hockey and Ice Skating championships take place. In India the Ice Hockey Association had held the Asian Championship out of the 7 winter federations such as Skiing, Biathlon, Ice Hockey, Ice Skating, Curling, Luge and Bobsleigh. No world association/ federation was prepared to give money to WGFI for the promotion of winter games as they were not sure that the money they gave would be used for the specific game. Narinder decided to bifurcate WGFI into independent small federations like Ice Hockey, Biathlon, Ice Skating, etc. That had the desired effect and all these associations are doing fine as far as finances are concerned.

Amongst all the winter games, the most expensive game is Ice Hockey. Apart from infrastructure, the team size is 24 and whenever a competition takes place abroad, India has to send the whole team there. Luckily, they had the Asian Ice Hockey Championship in Dehradun that was a great success. Bull Kumar remembers when he was the President of WGFI, he was sent to Lily Homar Winter Olympics and the Indian flag did not fly as they did not have even one athlete who could qualify for the Olympics. In fact, Ashwini Kumar, the then an I.O.C. member and the famous DGP. of the BSF was also there with the Indian Ambassador. They hung their heads in shame as they could not train even one athlete to take part in the games. They worked very hard and got sponsorships from the corporate world and foreign companies to train potential athletes.

Finally, when he left WGFI, there were 4 athletes in the Olympics competing for India -- two boys and a girl in alpine skiing and one in luge. It was unfortunate that instead of making progress, the winter games have declined. In the last Olympics, only 3 athletes participated from our country. Narinder doubts that until some pragmatic President takes over WGFI, things will improve. However, India was on the verge of making some sort of record in the Olympics as Shiva Keshwan, our luge athlete might become the only athlete in the world for taking part in the maximum number of Olympics.

It is a pity that due to politics in sports Indian Olympic Association was derecognized by International Olympic Council. Narinder is glad that during the Sochi Winter Olympics it was recognized and the Indian team that took part under the Olympic flag in the opening ceremony carried the Tri-Colour during the closing ceremony.

Narinder was indeed surprised to receive the MacGregor Medal 30 years after he had traversed the longest glacier in the Himalayas from snout to Indira Col. On the 13th of April 2014 that marked the 30th Anniversary of the Siachen War.

EPILOUGE

According to the history of the National Defence Academy, Colonel Bull Kumar (Retd) was perhaps the most decorated soldier of the 3 services – the Indian Army, the Indian Navy and the Indian Air Force. Here are the lists of awards that he has been awarded so far-

1963: Ati Vishist Seva Medal (AVSM) – That award is normally given to Brigadiers and above. But it was specially awarded to Captain Kumar for claiming Barahoti from China for India.

1966: Padma Shri – For arranging logistics support for nine a top Everest.

1966: Arjuna Award – All the members of 1965 Everest Expedition got Arjuna Awards.

1966: IMF Gold Medal – For overall contribution to Indian Mountaineering. Only 5 members in India got that medal before Captain Bull Kumar including Tenzing Norgay.

1977: Param Vishist Seva Medal (PVSM) – For being the leader of the Kanchenjunga Army Expedition from the North East Spur. In 1932 a German expedition led by Paul Bauer from the same route missed the summit by 3000 feet, but was given an Olympic Gold Medal for his great attempt.

1981: Kirti Chakra – For claiming Siachen Glacier - the largest glacier in the world from Pakistan for India.

1994: Most Innovative Tour Operator awarded by the Ministry of Tourism, Government of India.

1999: The Golden Paddle award from the Tourism Minister of Uttar Pradesh for promoting rafting in India.

2004: Lifetime Achievement Award awarded by Chief Minister J&K for promoting skiing in J&K.

2010: MacGregor Medal – For reconnaissance of Siachen Glacier that led to Operation Meghdoot.

2013: Lifetime Achievement Award awarded by the Chandigarh Tourism for promoting of adventure tourism in the Northern India.

Index

O

P

R

S